John Hunter, Joseph Banks

Observations on the Structure and Oeconomy of Whales

By John Hunter, Esq. F. R. S.; Communicated by Sir Joseph Banks, Bart. P.

R. S.

John Hunter, Joseph Banks

Observations on the Structure and Oeconomy of Whales
By John Hunter, Esq. F. R. S.; Communicated by Sir Joseph Banks, Bart. P. R. S.

ISBN/EAN: 9783337330125

Printed in Europe, USA, Canada, Australia, Japan

Cover: Foto ©berggeist007 / pixelio.de

More available books at **www.hansebooks.com**

XXXVIII. *Obfervations on the Structure and Oeconomy of* Whales.
By John Hunter, *Efq. F. R. S.; communicated by Sir* Jofeph
Banks, *Bart. P. R. S.*

Read June 28, 1787.

THE animals which inhabit the fea are much lefs known to us than thofe found upon land; and the œconomy of thofe with which we are beft acquainted is much lefs underftood: we are, therefore, too often obliged to reafon from analogy where information fails; which muft probably ever continue to be the cafe, from our unfitnefs to purfue our refearches in the unfathomable waters.

This unfitnefs does not arife from that part of our œconomy on which life and its functions depend; for the tribe of animals which is to be the fubject of this Paper, has, in that refpect, the fame œconomy as man, but from a difference in the mechanifm by which our progreffive motion is produced.

The anatomy of the larger marine animals, when they are procured in a proper ftate, can be as well afcertained as that of any others; dead ftructure being readily inveftigated. But even fuch opportunities too feldom occur, becaufe thofe animals are only to be found in diftant feas, which no one explores in purfuit of natural hiftory; neither can they be brought to us alive from thence, which prevents our receiving their bodies in a
ftate

ſtate fit for diſſcction.   As they cannot live in air, we are unable to procure them alive.

Some of theſe aquatic animals yielding ſubſtances which have become articles of traffic, and in quantity ſufficient to render them valuable as objects of profit, are ſought after for that purpoſe; but gain being the primary view, the reſearches of the Naturaliſt are only conſidered as ſecondary points, if conſidered at all.   At the beſt, our opportunities of examining ſuch animals do not often occur till the parts are in ſuch a ſtate as to defeat the purpoſes of accurate enquiry, and even theſe occaſions are ſo rare as to prevent our being able to ſupply, by a ſecond diſſection, what was deficient in a firſt.   The parts of ſuch animals being formed on ſo large a ſcale, is another cauſe which prevents any great degree of accuracy in their examination ; more eſpecially when it is conſidered, how very inconvenient for accurate diſſections are barges, open fields, and ſuch places as are fit to receive animals, or parts, of ſuch vaſt bulk.

As the opportunities of aſcertaining the anatomical ſtructure of large marine animals are generally accidental, I have availed myſelf, as much as poſſible, of all that have occurred ; and, anxious to get more extenſive information, engaged a Surgeon, at a conſiderable expence, to make a voyage to Greenland, in one of the ſhips employed in the whale fiſhery, and furniſhed him with ſuch neceſſaries as I thought might be requiſite for examining and preſerving the more intereſting parts, and with inſtructions for making general obſervations ; but the only return I received for this expence was a piece of whale's ſkin, with ſome ſmall animals ſticking upon it.   From the opportunities which I have had of examining different animals of this order, I have gained a tolerably accurate idea of the

anatomical

anatomical ftruɛture of fome genera, and fuch a knowledge of the ftruɛture of particular parts of fome others, as to enable me to afcertain the principles of their œconomy.

Thofe which I have had opportunities of examining were the following:

Of the Delphinus Phocæna, or Porpoife, I have had feveral, both male and female.

Of the Grampus I have had two; one of them (Tab. XVI.) twenty-four feet long, the belly of a white colour, which terminated at once, the fides and back being black; the other (Tab. XVII.) about eighteen feet long, the belly white, but lefs fo than in the former, and fhaded off into the dark colour of the back.

Of the Delphinus Delphis, or Bottle-nofe Whale (Tab. XVIII.) I had one fent to me by Mr. JENNER, Surgeon, at Berkeley. It was about eleven feet long. I have alfo had one twenty-one feet long, refembling this laft in the fhape of the head, but of a different genus, having only two teeth in the lower jaw (Tab. XIX.); the belly was white, fhaded off into the dark colour of the back. This fpecies is defcribed by DALE, in his Antiquities of Harwich. The one which I examined muft have been young; for I have a fkull of the fame kind, nearly three times as large, which muft have belonged to an animal thirty or forty feet long.

Of the Balæna roftrata of FABRICIUS, I had one, feventeen feet long (Tab. XX.).

The Balæna Myfticetus, or large Whalebone Whale, the Phyfeter Macrocephalus, or Spermaceti Whale, and the Monodon Monoceros, or Narwhale, have alfo fallen under my infpeɛtion. Some of thefe I have had opportunities of examining with accuracy; while others I have only examined in

part,

part, the animals having been too long kept before I procured them, to admit of more than a very fuperficial infpection.

From thefe circumftances it will be readily fuppofed, that an accurate defcription of all the different fpecies is not to be expected; but having acquired a general knowledge of the whole tribe, from the different fpecies which have come under my examination, I have been enabled to form a tolerable idea, even of parts which I have only had the opportunity of feeing in a very curfory way.

General obfervation would lead us to believe, that the whole of this tribe conftitutes one order of animals, which Naturalifts have fubdivided into genera and fpecies; but a deficiency in the knowledge of their œconomy has prevented them from making thefe divifions with fufficient accuracy; and this is not furprifing, fince the genera and fpecies are ftill in fome meafure undetermined even in animals with which we are better acquainted.

The animals of this order are in fize the largeft known, and probably, therefore, the feweft in number of all that live in water. Size, I believe, in thofe animals who feed upon others, is in an inverfe proportion to the number of the fmaller; but, I believe, this tribe varies more in that refpect than any we know, viewing it from the Whalebone Whale, which is feventy or eighty feet long, to the Porpoife that is five or fix: however, if they differ as much among themfelves as the Salmon does from the Sprat, there is not that comparative difference in fize that would at firft appear. The Whalebone Whale is, I believe, the largeft; the Spermaceti Whale the next in fize (the one which I examined, although not full grown, was about fixty feet long); the Grampus, which is an extenfive genus, is

probably

probably from twenty to fifty feet long; under this denomination there is a number of species.

From my want of knowledge of the different genera of this tribe of animals, an incorrectness in the application of the anatomical account to the proper genus may be the consequence; for when they are of a certain size, they are brought to us as Porpoises; when larger, they are called Grampus, or Fin-fish. A tolerably correct anatomical description of each species, with an accurate Drawing of the external form, would lead us to a knowledge of the different genera, and the species in each; and, in order to forward so useful a work, I propose, at some future period, to lay before the Society descriptions and drawings of those which have come under my own observation.

From some circumstances in their digestive organs we should be led to suppose, that they were nearly allied to each other; and that there was not the same variety, in this respect, as in land animals.

In the description of this order of animals, I shall always keep in view their analogy to land animals, and to such as occasionally inhabit the water, as white Bears, Seals, Manatees, &c. with the differences that occur. This mode of referring them to other animals, better known, will assist the mind in understanding the present subject. It is not, however, intended in this Paper to give a particular account of the structure of all the animals of this order, which I have had an opportunity of examining: I propose, at present, chiefly to confine myself to general principles, giving the great outlines as far as I am acquainted with them, minuteness being only necessary in the investigation of particular parts.

In my account I ſhall pay ſome attention to the relations of men who have given faɛts without knowing their cauſes, whenever I find that ſuch faɛts can be explained upon true principles of the animal œconomy, but no further.

This order of animals has nothing peculiar to fiſh, except living in the ſame element, and being endowed with the ſame powers of progreſſive motion as thoſe fiſh that are intended to move with a conſiderable velocity: for I believe, that all that come to the ſurface of the water (which this order of animals muſt do) have conſiderable progreſſive motion; and this reaſoning we may apply to birds; for thoſe which ſoar very high have the greateſt progreſſive motion.

Although inhabitants of the waters, they belong to the ſame claſs as quadrupeds, breathing air, being furniſhed with lungs, and all the other parts peculiar to the œconomy of that claſs, and having warm blood; for we may make this general remark, that in the different claſſes of animals there is never any mixture of thoſe parts which are eſſential to life, nor in their different modes of ſenſation.

I ſhall divide what is called the œconomy of an animal, firſt, into thoſe parts and aɛtions which reſpeɛt its internal funɛtions, and on which life immediately depends, as growth, waſte, repair, ſhifting or changing of parts, &c. the organs of reſpiration and ſecretion, in which we may include the powers of propagating the ſpecies.

Secondly, into thoſe parts and aɛtions which reſpeɛt external objeɛts, and which are variouſly conſtruɛted, according to the kind of matter with which they are to be conneɛted, whence they vary more than thoſe of the firſt diviſion. Theſe are the parts for progreſſive motion, the organs of ſenſe

and

and the organs of digeftion; all which either act, or are acted upon, by external matter.

This variation from external caufes in many inftances in-fluences the fhape of the whole, or particular parts, even giving a peculiar form to fome which belong to the firft order of actions, as the heart, which in this tribe, in the Seal, Otter, &c. is flattened, becaufe the cheft is flattened for the purpofe of fwimming. The contents of the abdomen are not only adapted to the external form; but their direction in the cavity is, in fome inftances, regulated by it. The anterior extremity, or fin, although formed of diftinct parts, in fome degree fimilar to the anterior extremities of fome quadrupeds, being com-pofed of fimilar bones placed nearly in the fame manner, yet are fo formed and arranged as to fit them for progreffive motion in the water only.

The external form of this order of animals is fuch as fits them for dividing the water in progreffive motion, and gives them power to produce that motion in the fame manner as thofe fifh which move with a confiderable velocity. On account of their inhabiting the water, their external form is more uniform than in animals of the fame clafs which live upon land, the furface of the earth on which the progreffive motion of the quadruped is to be performed being various and irregular, while the water is always the fame.

The form of the head or anterior part of this order of animals is commonly a cone, or an inclined plane, except in the Sperma-ceti Whale, in which it terminates in a blunt furface. This form of head increafes the furface of contact to the fame volume of water which it removes, leffens the preffure, and is better calculated to bear the refiftance of the water through which the animal is to pafs; probably, on this account, the head

H h h 2   is

is larger than in quadrupeds, having more the proportion obferved in fifh, and fwelling out laterally at the
articulation of the lower jaw : this may probably be for
the better catching their prey, as they have no motion of
the head on the body; and this diftance between the articulations of the jaw is fomewhat fimilar to the Swallow,
Goat-fucker, Bat, &c. which may alfo be accounted for,
from their catching their food in the fame manner as fifh; and
this is rendered ftill more probable, fince the form of the
mouth varies according as they have or have not teeth.   There
is, however, in the Whale tribe more variety in the form of the
head than of any other part, as in the Whalebone, Bottle-
nofe, and Spermaceti Whales; though in this laft it appears
to owe its fhape, in fome fort, to the vaft quantity of fpermaceti
lodged there, and not to be formed merely for the catching of its
prey.   From the mode of their progreffive motion, they have
not the connection between the head and body, that is called
the neck, as that would have produced an inequality inconvenient to progreffive motion.

The body behind the fins or fhoulders diminifhes gradually
to the fpreading of the tail; but the part beyond the opening of the anus is to be confidered as tail, although to appearance it is a continuation of the body.   The body itfelf is
flattened laterally; and, I believe, the back is much fharper
than the belly.

The projecting part, or tail, contains the power that produces progreffive motion, and moves the broad termination,
the motion of which is fimilar to that of an oar in fculling a
boat; it fuperfedes the neceffity of pofterior extremities, and
allows of the proper fhape for fwimming; that the form may
be preferved as much as poffible, we find that all the projecting

6   parts,

parts, found in land animals of the same class, are either in-tirely wanting, as the external ear; are placed internally, as the testicles; or are spread along under the skin, as the udder.

The tail is flattened horizontally, which is contrary to that of fish, this position of tail giving the direction to the ani-mal in the progressive motion of the body. I shall not pursue this circumstance further than to apply it to those purposes in the animal œconomy, for which this particular direction is intended.

The two lateral fins, which are analogous to the anterior extremities in the quadruped, are commonly small, varying however in size, and seem to serve as a kind of oars.

To ascertain the use of the *fin* on the back is probably not so easy, as the large Whalebone and Spermaceti Whales have it not; one should otherwise conceive it intended to preserve the animal from turning.

I believe, like most animals, they are of a lighter colour on their belly than on their back : in some they are intirely white on the belly ; and this white colour begins by a regular deter-mined line, as in the Grampus, Piked Whale, &c. : in others, the white on the belly is gradually shaded into the dark colour of the back, as in the Porpoise. I have been informed, that some of them are pied upwards and downwards, or have the divisions of colour in a contrary direction.

The element in which they live renders certain parts which are of importance in other animals useless in them, gives to some parts a different action, and renders others of less account.

The puncta lachrymalia with the appendages, as the sac and duct, are in them unnecessary ; and the secretion from the lachry-mal gland is not water, but mucus, as it also is in the Turtle ;

and

and we may fuppofe only in fmall quantity, the gland itf
being fmall.

The urinary bladder is fmaller than in quadrupeds; and i
deed there is not any apparent reafon why Whales fhould ha
one at all.

The tongue is flat, and but little projecting, as they neitl
have voice, nor require much action of this part, in applyi;
the food between the teeth for the purpofe of maftication,
deglutition, being nearly fimilar to fifh in this refpect, as w
as in their progreffive motion.

In fome particulars they differ as much from one another
any two genera of quadrupeds I am acquainted with.

The larynx, fize of trachea, and number of ribs, dif
exceedingly. The cæcum is only found in fome of the
The teeth in fome are wanting. The blow-holes are two
number in many, in others only one. The whalebone a
fpermaceti are peculiar to particular genera: all which conf
tute great variations. In other refpects we find an uniformit
which would appear to be independent of their living a
moving only in the water, as in the ftomach, liver, parts
generation of both fexes, and in the kidneys: in thefe 1
however, I believe, it depends in fome degree upon their fitt
tion, although it is extended to other animals, the caufe
which I do not underftand.

All animals have, I believe, a fmell peculiar to themfelve
how far this is connected with the other diftinctions, I do r
know, our organs not being able to diftinguifh with fufficie
accuracy.

The fmell of animals of this tribe is the fame with that
the Seal, but not fo ftrong, a kind of four fmell, which t

Seal has while alive; the oil has the fame fmell with that of the Salmon, Herring, Sprat, &c.

The obfervations refpecting the weight of the flefh of animals that fwim, which I publifhed in my obfervations on the œconomy of certain parts of animals, are applicable to thefe alfo; for the flefh in this tribe is rather heavier than beef; two portions of mufcle of the fame fhape, one from the pfoas mufcle of the Whale, the other of an ox, when weighed in air, were both exactly 502 grains; but, weighed in water, the portion of the Whale was four grains heavier than the other. It is probable, therefore, that the neceffary equilibrium between the water and the animal is produced by the oil, in addition to which the principal action of the tail is fuch as tends either to raife them, or keep them fufpended in the water, according to the degree of force with which it acts.

From the tail being horizontal, the motion of the animal, when impelled by it, is up and down: two advantages are gained by this, it gives the neceffary opportunity of breathing, and elevates them in the water; for every motion of the tail tends, as I faid before, to raife the animal: and that this may be effected, the greateft motion of the tail is downwards, thofe mufcles being very large, making two ridges in the abdomen; this motion of the tail raifes the anterior extremity, which always tends to keep the body fufpended in the water.

Of the Bones.

The bones alone, in many animals, when properly united into what is called the fkeleton, give the general fhape and character of the animal. Thus a quadruped is diftinguifhed from a bird, and even one quadruped from another, it only
requiring

requiring a fkin to be thrown over the fkeleton to make the fpecies known; but this is not fo decidedly the cafe with this order of animals, for the fkeleton in them does not give us the true fhape. An immenfe head, a fmall neck, few ribs, and in many a fhort fternum, and no pelvis, with a long fpine, terminating in a point, require more than a fkin being laid over them to give the regular and charaćteriftic form of the animal.

The bones of the anterior extremity give no idea of the fhape of a fin, the form of which depends wholly upon its covering. The different parts of the fkeleton, are fo inclofed, and the fpaces between the projećting parts are fo filled up, as to be altogether concealed, giving the animal externally an uniform and elegant form, refembling an infećt enveloped in its chryfalis coat.

The bones of the head are in general fo large, as to render the cavity which contains the brain but a fmall part of the whole; while, in the human fpecies, and in birds, this cavity conftitutes the principal bulk of the head. This is, perhaps, moft remarkable in the Spermaceti Whale; for on a general view of the bones of the head, it is impoffible to determine where the cavity of the fkull lies, till led to it by the foramen magnum occipitale. The fame remark is applicable to the large Whalebone and Bottle-nofe Whale; but in the Porpoife, where the brain is larger in proportion to the fize of the animal, the fkull makes the principal part of the head.

Some of the bones in one genus differ from thofe of another. The lower jaw is an inftance of this. In the Spermaceti and Bottle-nofe Whales, the Grampus, and the Porpoife, the lower jaws, efpecially at the pofterior ends, refemble each other;

other; but in both the large and small Whalebone Whales, the shape differs considerably. The number of some particular bones varies likewise very much.

The Piked Whale has seven vertebræ in the neck, twelve which may be reckoned to the back, and twenty-seven to the 'tail, making forty-six in the whole.

In the porpoise there are five cervical vertebræ, and one common to the neck and back, fourteen proper to the back, and 30 to the tail, making in the whole fifty-one.

The small Bottle-nose Whale, caught near Berkeley, in the number of cervical vertebræ resembled the Porpoise; it had seventeen in the back, and thirty-seven in the tail, in all sixty.

In the Porpoise, four of the vertebræ of the neck are anchylosed; and in every animal of this order, which I have examined, the atlas is by much the thickest, and seems to be made up of two joined together, for the second cervical nerve passes through a foramen in this vertebra. There is no articulation for rotatory motion between the first and second vertebræ of the neck.

The small Bottle-nose Whale had eighteen ribs on each side, the Porpoise sixteen. The ends of the ribs that have two articulations, in the whole of this tribe, I believe, are articulated with the body of the vertebræ above, and with the transverse processes below, by the angles; so that there is one vertebra common to the neck and back. In the large Whalebone Whale the first rib is bifurcated, and consequently articulated to two vertebræ.

The sternum is very flat in the Piked Whale; it is only one very short bone; and in the Porpoise it is a good deal longer. In the small Bottle-nose it is composed of three bones, and is of ſome length. In the Piked Whale the firſt rib, and in the Porpoiſe the three firſt, are articulated with the ſternum.

As a contraction, correſponding to the neck in quadrupeds, would have been improper in this-order of animals, the vertebræ of the neck are thin, to make the diſtance between the head and ſhoulders as ſhort as poſſible, and in the ſmall Bottle-noſe Whale are only ſix in number.

The ſtructure of the bones is ſimilar to that of the bones of quadrupeds; they are compoſed of an animal ſubſtance, and an earth that is not animal: theſe ſeem only to be mechanically mixed, or rather the earth thrown into the interſtices of the animal part. In the bones of fiſhes this does not ſeem to be the caſe, the earth in many fiſh being ſo united with the animal part, as to render the whole tranſparent, which is not the caſe when the animal part is removed by ſteeping the bone in cauſtic alkali: nor is the animal part ſo tranſparent when deprived of the earth. The bones are leſs compact than thoſe of quadrupeds that are ſimilar to them.

Their form ſomewhat reſembles what takes place in the quadruped, at leaſt in thoſe whoſe uſes are ſimilar, as the vertebræ, ribs, and bones of the anterior extremities have their articulations in part alike, although not in all of them. The articulation of the lower jaw, of the carpus, metacarpus, and fingers, are exceptions. The articulation of the lower jaw is not by ſimple contact either ſingle or double, joined by a capſular ligament, as in the quadruped; but by a very thick intermediate ſubſtance of the ligamentous kind, ſo interwoven that its parts move on each other, in the interſtices of which is an oil. This thick matted ſubſtance may anſwer the ſame purpoſe as the double joint in the quadruped.

The

The two fins are analogous to the anterior extremities of the quadruped, and are alfo fomewhat fimilar in conftruction. A fin is compofed of a fcapula, os humeri, ulna, radius, carpus, and metacarpus, in which laft may be included the fingers, becaufe the number of bones are thofe which might be called fingers, although they are not feparated, but included in one general covering with the metacarpus. They have nothing analogous to the thumb, and the number of bones in each is different; in the fore-finger there are five bones, in the middle and ring-finger feven, and in the little finger four. The articulation of the carpus, metacarpus, and fingers, is different from that of the quadruped, not being by capfular ligament, but by intermediate cartilages connected to each bone. Thefe cartilages between the different bones of the fingers are of confiderable length, being nearly equal to one-half of that of the bone; and this conftruction of the parts gives firmnefs, with fome degree of pliability, to the whole.

As this order of animals cannot be faid to have a pelvis, they of courfe have no os facrum, and therefore the vertebræ are continued on to the end of the tail; but with no diftinction between thofe of the loins and tail. But as thofe vertebræ alone would not have had fufficient furface to give rife to the mufcles requifite to the motion of the tail, there are bones added to the fore-part of fome of the firft vertebræ of the tail, fimilar to the fpinal proceffes on the pofterior furface.

From all thefe obfervations we may infer, that the ftructure, formation, arrangement, and the union of the bones, which compofe the forms of parts in this order of animals, are much upon the fame principle as in quadrupeds.

The flefh or mufcles of this order of animals is red, refembling that of moft quadrupeds, perhaps more like that of the

Bull

Bull or Horſe than any other animal: ſome of it is very firm; and about the breaſt and belly it is mixed with tendon.

Although the body and tail is compoſed of a ſeries of bones connected together and moved as in fiſh, yet it has its movements produced by long muſcles, with long tendons, which renders the body thicker, while the tail at its ſtem is ſmaller than that of any other ſwimmer, whoſe principal motion is the ſame. Why this mode of applying the moving powers ſhould not have been uſed in fiſh, is probably not ſo eaſily anſwered; but in fiſh the muſcles of the body are of nearly the ſame length as the vertebræ.

The depreſſor muſcles of the tail, which are ſimilar in ſituation to the pſoæ, make two very large ridges on the lower part of the cavity of the belly, riſing much higher than the ſpine, and the lower part of the aorta paſſes between them.

Theſe two large muſcles, inſtead of being inſerted into two extremities as in the quadruped, go to the tail, which may be conſidered in this order of animals as the two poſterior extremities united into one.

Their muſcles, a very ſhort time after death, loſe their fibrous ſtructure, become as uniform in texture as clay or dough, and even ſofter. This change is not from putrefaction, as they continue to be free from any offenſive ſmell, and is moſt remarkable in the pſoæ muſcles, and thoſe of the back.

Of the Conſtruction of the Tail.

The mode in which the tail is conſtructed is, perhaps, as beautiful, as to the mechaniſm, as any part of the animal. It is wholly compoſed of three layers of tendinous fibres, covered by the common cutis and cuticle: two of theſe layers

4   are

are external, the other internal. The direction of the fibres of the external layers is the same as in the tail, forming a stratum about one-third of an inch thick; but varying, in this respect, as the tail is thicker or thinner. The middle layer is composed entirely of tendinous fibres, passing directly acrofs, between the two external ones above described, their length being in proportion to the thickness of the tail; a structure which gives amazing strength to this part.

The substance of the tail is so firm and compact, that the vessels retain their dilated state, even when cut acrofs; and this section consists of a large vessel surrounded by as many small ones as can come in contact with its external surface; which of these are arteries, and which veins, I do not know.

The fins are merely covered with a strong condensed adipose membrane.

Of the Fat.

The fat of this order of animals, except the spermaceti, is what we generally term oil. It does not coagulate in our atmosphere, and is probably the most fluid of animal fats; but the fat of every different order of animals has not a peculiar degree of solidity, some having it in the same state, as the Horse and Bird. What I believe approaches nearest to spermaceti, is the fat of ruminating animals, called tallow.

The fat is differently situated in different orders of animals; probably for particular purposes, at least in some we can assign a final intention. In the animals, which are the subject of the present Paper, it is found principally on the outside of the mufcles, immediately under the skin, and is in confiderable quantity. It is rarely to be met with in the interstices of the muscles, or in any of the cavities, such as the abdomen or about the heart.

In

In animals of the fame clafs living on land, the fat is more diffufed: it is fituated, more efpecially when old, in the interftices of mufcles, even between the fafciculi of mufcular fibres, and is attached to many of the vifcera ; but many parts are free from fat, unlefs when difeafed, as the penis, fcrotum, tefticle, eyelid, liver, lungs, brain, fpleen, &c.

In fifh its fituation is rather particular, and is moft commonly in two modes ; in the one, diffufed through the whole body of the fifh, as in the Salmon, Herring, Pilchard, Sprat, &c. ; in the other, it is found in the liver only, as in all of the Ray kind, Cod, and in all thofe called White-fifh, there being none in any other part of the body *. The fat of fifh appears to be diffufed through the fubftance of the parts which contain it, but is probably in diftinct cells. In fome of thefe fifh, where it is diffufed over the whole body, it is more in fome parts than others, as on the belly of the Salmon, where it is in larger quantity.

The fat is differently inclofed in different orders of animals. In the quadruped, thofe of the Seal kind excepted, in the bird, amphibia, and in fome fifh, it is contained in loofe cellular membrane, as if in bags, compofed of fmaller ones, by which means the larger admit of motion on one another, and on their connecting parts ; which motion is in a greater or lefs degree, as is proper or ufeful. Where motion could anfwer no purpofe, as in the bones, it is confined in ftill fmaller cells. The fat is in a lefs degree in the foles of the feet, palms of the hands, and in the breafts of many animals. In this order of animals and the Seal kind, as far as I yet know, it is difpofed of in two ways ; the fmall quantity found in the cavities of the body,

---

* The Sturgeon is, however, an exception, having its fat in particular fituations, and in the interftices of parts, as in other animals.

and

and interstices of parts, is in general disposed in the same way as
in quadrupeds; but the external, which includes the principal
part, is inclosed in a reticular membrane, apparently composed of
fibres passing in all directions, which seem to confine its extent,
allowing it little or no motion on itself, the whole, when diftended, forming almost a solid body. This, however, is not
always the case in every part of animals of this order; for
under the head, or what may be rather called neck, of the Bottle-
nose, the fat is confined in larger cells, admitting of motion.
This reticular membrane is very fine in some, and very strong
and coarse in others, and even varies in different parts of the
same animal. It is fine in the Porpoise, Spermaceti, and
large Whalebone Whale; coarse in the Grampus and small
Whalebone Whale *: in all of them it is finest on the body,
becoming coarser towards the tail, which is composed of fibres
without any fat : which is also the case in the covering of the
fins. This reticular net-work in the Seal is very coarse; and
in those which are not fat, when it collapses, it looks almost
like a fine net with small meshes. This structure confines the
animal to a determined shape, whereas in quadrupeds fat when
in great quantity destroys all shape.

The fat differs in consistence in different animals, and in different parts of the same animal, in which its situation is various.
In quadrupeds, some have the external fat softer than the internal;
and that inclosed in bones is softest nearer to their extremities.
Ruminating animals have that species of fat called tallow, and in
their bones they have either hard fat or marrow, or fluid fat called
Neat's-foot oil. In this order of animals, the internal fat is the
least fluid, and is nearly of the consistence of Hog's-lard; the

---

* Where it is fine, it yields the largest quantity of oil, and requires the least
boiling.

I   external

external is the common train oil; but the Spermaceti Whale differs from every other animal I have examined, having the two kinds of fat juft mentioned, and another which is totally different, called fpermaceti, of which I fhall give a particular account.

What is called fpermaceti is found every where in the body in fmall quantity, mixed with the common fat of the animal, bearing a very fmall proportion to the other fat. In the head it is the reverfe, for there the quantity of fpermaceti is large when compared to that of the oil, although they are mixed, as in the other parts of the body.

As the fpermaceti is found in the largeft quantity in the head, and in what would appear on a flight view to be the cavity of the fkull, from a peculiarity in the fhape of that bone, it has been imagined by fome to be the brain.

Thefe two kinds of fat in the head are contained in cells, or cellular membrane, in the fame manner as the fat in other animals; but befides the common cells there are larger ones, or ligamentous partitions going acrofs, the better to fupport the vaft load of oil, of which the bulk of the head is principally made up.

There are two places in the head where this oil lies; thefe are fituated along its upper and lower part: between them pafs the noftrils, and a vaft number of tendons going to the nofe and different parts of the head.

The pureft fpermaceti is contained in the fmalleft and leaft ligamentous cells: it lies above the noftril, all along the upper part of the head, immediately under the fkin, and common adipofe membrane. Thefe cells refemble thofe which contain the common fat in the other parts of the body neareft the fkin. That which lies above the roof of the mouth, or between it

and

and the noftril, is more intermixed with a ligamentous cellular membrane, and lies in chambers whofe partitions are perpendicular. Thefe chambers are fmaller the nearer to the nofe, becoming larger and larger towards the back part of the head, where the fpermaceti is more pure.

This fpermaceti, when extracted cold, has a good deal the appearance of the internal ftructure of a water melon, and is found in rather folid lumps.

About the nofe, or anterior part of the noftril, I difcovered a great many veffels, having the appearance of a plexus of veins, fome as large as a finger. On examining them, I found they were loaded with the fpermaceti and oil; and that fome had correfponding arteries. They were moft probably lymphatics; therefore I fhould fuppofe, that their contents had been abforbed from the cells of the head. We may the more readily fuppofe this, from finding many of the cells, or chambers, almoft empty; and as we may reafonably believe that this animal had been fome time out of the feas in which it could procure proper food, it had perhaps lived on the fuperabundance of oil.

The folid maffes are what are brought home in cafks for fpermaceti.

I found, by boiling this fubftance, that I could eafily extract the fpermaceti and oil which floated on the top from the cellular membrane. When I fkimmed off the oily part, and let it ftand to cool, I found that the fpermaceti cryftallifed, and the whole became folid; and by laying this cake upon any fpongy fubftance, as chalk, or on a hollow body, the oil drained all off, leaving the fpermaceti pure and white. Thefe cryftals were only attached to each other by edges, forming a fpongy mafs; and by melting this pure fpermaceti, and allowing it to cryftallife, it was reduced in appearance to half its bulk, the cryftals being fmaller, and more blended, confequently lefs diftinct.

The fpermaceti mixes readily with other oils, while it is in a fluid ftate, but feparates or cryftallifes whenever it is cooled to a certain degree; like two different falts being diffolved in water, one of which will cryftallife with a lefs degree of evaporation than the other; or, if the water is warm, and fully faturated, one of the falts will cryftallife fooner than the other, while the folution is cooling. I wanted to fee whether fpermaceti mixed equally well with the expreffed oils of vegetables when warm, and likewife feparated and cryftallifed when cold, and on trial there feemed to be no difference. When very much diluted with the oil, it is diffolved or melted by a much fmaller degree of heat than when alone; and this is the reafon, perhaps, that it is in a fluid ftate in the living body.

If the quantity of fpermaceti is fmall in proportion to the other oil, it is, perhaps, nearly in that proportion longer in cryftallifing; and when it does cryftallife, the cryftals are much fmaller than thofe that are formed where the proportion of fpermaceti is greater. From the flownefs with which the fpermaceti cryftallifes when much diluted with its oil, from a confiderable quantity being to be obtained in that way, and from its continuing for years to cryftallife, one would be induced to think, that perhaps the oil itfelf is converted into fpermaceti.

It is moft likely, that if we could difcover the exact form of the different cryftals of oils, we fhould thence be able to afcertain both the different forts of vegetable oils, expreffed and effential, and the different forts of animal oils, much better than by any other means; in the fame manner as we know falts by the forms into which they fhoot.

6   The

The fpermaceti does not become rancid, or putrid, nearly fo foon as the other animal oils; which is moft probably owing to the fpermaceti being for the moft part in a folid ftate; and I fhould fuppofe, that few oils would become fo foon rancid as they do, if they were always preferved in that degree of cold which rendered them folid: neither does this oil become fo foon putrid as the flefh of the animal; and therefore, although the oil in the cells appeared to be putrid before boiling, it was fweet when deprived of the cellular fubftance. The fpermaceti is rather heavier than the other oil.

In this animal then we find two forts of oil, befides the deeper feated fat, common to all of this clafs; one of which cryftallifes with a much lefs degree of cold than the other, and of courfe requires a greater degree of heat to melt it, and forms, perhaps, the largeft cryftals of any exprefled oil we know: yet the fluid oil of this animal will cryftallife in an extreme hard froft, much fooner than moft effential oils, though not fo foon as the exprefled oils of vegetables. Camphire, however, is an exception, fince it cryftallifes in our warmeft weather, and when melted with exprefled oil of vegetables, if the oil is too much faturated for that particular degree of cold, cryftallifes exactly like fpermaceti.

In the Ox the tallow, and what is called Neat's-foot oil, cryftallife in different degrees of cold. The tallow congeals with rather lefs cold than the fpermaceti; but the other oil is fimilar to what is called the train oil in the Whale.

I have endeavoured to difcover the form of the cryftals of different forts of oil; but could never determine exactly what that was, becaufe I could never find any of the cryftals fingle, and by being always united, the natural form was not diftinct.

It

It is the adipoſe covering from all of the Whale kind that is brought home in ſquare pieces, called flitches, and which, by being boiled, yields the oil on expreſſion, leaving the cellular membrane. When theſe flitches have become in ſome degree putrid, there iſſues two ſorts of oil; the firſt is pure, the laſt ſeems incorporated with part of the animal ſubſtance, which has become eaſy of ſolution from its putridity, forming a kind of butter. It is unctuous to the touch, ropy, coagulates or becomes harder by cold, ſwims upon water, not being ſoluble in it; and the pure oil, ſeparating in the ſame manner from this, ſwims above all.

What remains, after all the oil is extracted, retains a good deal of its form, is almoſt wholly convertible into glue, and is ſold for that purpoſe.

The cellular, or rather what ſhould be called the uniting membrane in this order of animals, is ſimilar to that in the quadruped; we find it uniting muſcle to muſcle, and muſcle to bone, for their eaſy motion on one another.

The cellular membrane, which is the receptacle for the oil near the ſurface of the body is in general very different from that in the quadruped, as has been already obſerved.

Of the Skin.

The covering of this order of animals conſiſts of a cuticle and cutis.

The cuticle is ſomewhat ſimilar to that on the ſole of the foot in the human ſpecies, and appears to be made up of a number of layers, which ſeparate by ſlight putrefaction; but this I ſuſpect ariſes in ſome degree from there being a ſucceſſion of cuticles formed. It has no degree of elaſticity or toughneſs,

4   but

but tears eafily; nor do its fibres appear to have any particular direction. The internal ftratum is tough and thick, and in the Spermaceti Whale its internal furface, when feparated from the cutis, is juft like coarfe velvet, each pile ftanding firm in its place; but this is not fo diftinguifhable in fome of the others, although it appears rough from the innumerable perforations.

It is the cuticle that gives the colour to the animal; and in parts that are dark, I think, I have feen a dirty coloured fubftance wafhed away in the feparation of the cuticle from the cutis, which muft be a kind of rete mucofum.

The cutis in this tribe is extremely villous on its external furface, anfwering to the rough furface of the cuticle, and forming in fome parts fmall ridges, fimilar to thofe on the human fingers and toes. Thefe villi are foft and pliable; they float in water, and each is longer or fhorter according to the fize of the animal. In the Spermaceti Whale they were about a quarter of an inch long; in the Grampus, Bottle-nofe and Piked Whales, much fhorter; in all, they are extremely vafcular.

The cutis feems to be the termination of the cellular membrane of the body more clofely united, having fmaller interftices, and becoming more compact. This alteration in the texture is fo fudden as to make an evident diftinction between what is folely connecting membrane, and fkin, and is moft evident in lean animals; for in the change from fat to lean, the fkin does not undergo an alteration equal to what takes place in the adipofe membrane, although it may be obferved, that the fkin itfelf is diminifhed in thicknefs. In fat animals the diftinction between fkin and cellular membrane is much lefs, the gradation from the one to the other feeming to be flower; for the cells of both membrane and fkin being loaded with fat, the whole
has

has more the appearance of one uniform ſubſtance. This uniformity of the adipoſe membrane and ſkin is moſt obſervable in the Whale, Seal, Hog, and the human ſpecies; and is not only viſible in the raw but in the dreſſed hides; for in dreſſed ſkins the external is much more compact in texture than the inner ſurface, and is in common very tough.

In ſome animals the cutis is extremely thick, and in ſome parts much more ſo than others : where very thick, it appears to be intended as a defence againſt the violence of their own ſpecies or other animals. In moſt quadrupeds it is muſcular, contracting by cold, and relaxing by heat. Many other ſtimulating ſubſtances make it contract; but cold is probably that ſtimulus by which it was intended to be generally affected.

The ſkin is extremely elaſtic in the greateſt number of quadrupeds, and in its contracted ſtate may be ſaid to be rather too ſmall for the body; by this elaſticity it adapts itſelf to the changes which are conſtantly taking place in the parts, and it is from the want of it, that it becomes too large in ſome old animals. In all animals it is more elaſtic in ſome parts than others, eſpecially in thoſe where there is the greateſt motion. How far theſe variations take place in the Whale I do not exactly know; but a looſe elaſtic ſkin in this tribe would appear to be improper as an univerſal covering, conſidering the progreſſive motion of the animal, and the medium in which it moves; therefore it appears to be kept always on the ſtretch, by the adipoſe membrane being loaded with fat, which does not allow the ſkin to recede when cut. It is, however, more elaſtic at the ſetting on of the eyelids, round the opening of the prepuce, the nipples, the ſetting on of the fins, and under the jaw, to allow of motion in thoſe parts;

and

and here there is more reticular, and lefs adipofe membrane.
But in the Piked Whale there is probably one of the moft
ftriking inftances of an elaftic cuticular contraction: for
though the whole fkin of the fore part of the neck and breaft' of
the animal, as far down as the middle of the belly, be extremely
elaftic ; yet to render it ftill more fo, it is ribbed longitudinally
like a ribbed ftocking, which gives an increafed lateral elafticity. Thefe ribs are, when contracted, about five-eighths of an
inch broad, covered with the common fkin of the animal; but in
the hollow part of the rib, it is of a fofter texture, with a thinner
cuticle. This part is poffeffed of the greateft elafticity; why
it fhould be fo elaftic is difficult to fay, as it covers the
thorax, which can never be increafed in fize; yet there muft
be fome peculiar circumftance in the œconomy of the fpecies
requiring this ftructure, which we as yet know nothing of.

The fkin is intended for various purpofes. It is the univer-
fal covering given for the defence of all kinds of animals; and
that it might anfwer this purpofe well, it is the feat of one of
the fenfes.

Of the Mode of catching their Food.

The mouths of animals are the firft parts to be confidered
refpecting nourifhment or food, and are fo much connected
with every thing relative to it, as not only to give good hints
whether the food is vegetable or animal, but alfo refpecting
the particular kind of either, efpecially of animal food. The
mouth not only receives the food, but is the immediate inftrument for catching it. As it is a compound inftrument in many
animals, having parts of various conftructions belonging to it,
I fhall at prefent confider it in this tribe no further than as
connected

connected with their mode of catching the food, and adapting and difpofing it for being fwallowed. It is probable, that thefe animals do not require either a divifion of the food, or a maftication of it in the mouth, but fwallow whatever they catch, whole; for we do not find any of them furnifhed with parts capable of producing either effect. The mouth in moft of this tribe is well adapted for catching the food; the jaws fpread as they go back, making the mouth proportionally wider than in many other animals.

There is a very great variety in the formation of the mouths of this tribe of animals, which we have many opportunities of knowing, from the head being often brought home when the other parts of the animal are rejected; a circumftance which frequently leaves us ignorant of the particular fpecies to which it belonged.

Some catch their food by means of teeth, which are in both jaws, as the Porpoife and Grampus; in others, they are only in one jaw, as in the Spermaceti Whale; and in the large Bottle-nofe Whale, defcribed by Dale, there are only two fmall teeth in the anterior part of the lower jaw. In the Narwhale only two tufks in the fore part of the upper jaw *; while in fome others there are none at all. In thofe which have teeth in both jaws, the number in each varies confiderably; the fmall Bottle-nofe had forty-fix in the upper, and fifty in the lower; and in the jaws of others there are only five or fix in each.

The teeth are not divifible into different claffes, as in quadrupeds; but are all pointed teeth, and are commonly a good deal fimilar. Each tooth is a double cone, one point being

* I call thefe tufks to diftinguifh them from common teeth. A tufk is that kind of tooth which has no bounds fet to its growth, excepting by abrafion, as the tufk of the Elephant, Boar, Sea-horfe, Manatee, &c,

faftened

faftened in the gum, the other projecting: they are, however, not all exactly of this fhape. In fome fpecies of Porpoife the fang is flattened, and thin at its extremity; in the Spermaceti Whale the body of the tooth is a little curved towards the back part of the mouth; this is alfo the cafe in fome others. The teeth are compofed of animal fubftance and earth, fimilar to the bony part of the teeth in quadrupeds. The upper teeth are commonly worn down upon the infide, the lower on the out-fide; this arifes from the upper jaw being in general the largeft.

The fituation of the teeth, when firft formed, and their progrefs afterwards, as far as I have been able to obferve, is very different in common from thofe of the quadruped. In the quadruped the teeth are formed in the jaw, almoft furrounded by the alveoli, or fockets, and rife in the jaw as they increafe in length; the covering of the alveoli being abforbed, the alveoli afterwards rife with the teeth, covering the whole fang; but in this tribe the teeth appear to form in the gum, upon the edge of the jaw, and they either fink in the jaw as they lengthen, or the alveoli rife to inclofe them: this laft is moft probable, fince the depth of the jaw is alfo increafed, fo that the teeth appear to fink deeper and deeper in the jaw. This formation is readily dif-covered in jaws not full grown; for the teeth increafe in number as the jaw lengthens, as in other animals. The pofterior part of the jaw becoming longer, the number of teeth in that part increafes, the fockets becoming fhallower and fhallower, and at laft being only a flight depreffion.

It would appear, that they do not fhed their teeth, nor have they new ones formed fimilar to the old, as is the cafe with moft other quadrupeds, and alfo with the Alligator. I have never been able to detect young teeth under the roots of the old ones; and indeed the fituation in which they are firft formed makes it in fome degree impoffible, if the young teeth follow the fame rule in growing with the original ones, as they probably do in moft animals.

If it is true, that the Whale tribe do not fhed their teeth, in what way are they fupplied with new ones, correfponding in fize with the increafed fize of the jaw? It would appear, that the jaw, as it increafes pofteriorly, decays at the fymphyfis, and while the growth is going on, there is a conftant fucceffion of new teeth, by which means the new-formed teeth are proportioned to the jaw. The fame mode of growth is evident in the Elephant, and in fome degree in many fifh; but in thefe laft the abforption of the jaw is from the whole of the outfide along where the teeth are placed. The depth of the alveoli feems to prove this, being fhallow at the back part of the jaw, and becoming deeper towards the middle, where they are the deepeft, the teeth there having come to the full fize. From this forwards they are again becoming fhallower, the teeth being fmaller, the fockets wafting, and at the fymphyfis there are hardly any fockets at all. This will make the exact number of teeth in any fpecies uncertain.

Some genera of this tribe have another mode of catching their food, and retaining it till fwallowed, which is by means of the fubftance called Whalebone. Of this there are two kinds known; one very large, probably from the largeft Whale yet difcovered; the other from a fmaller fpecies.

This whalebone, which is placed on the infide of the mouth, and attached to the upper jaw, is one of the moft fingular circumftances belonging to this fpecies, as they have moft other parts in common with quadrupeds. It is a fubftance, I believe, peculiar to the Whale, and of the fame nature as horn, which I fhall ufe as a term to exprefs what conftitutes

tutes hair, nails, claws, feathers, &c. it is wholly compofed of animal fubftance, and extremely elaftic *.

Whalebone confifts of thin plates of fome breadth, and in fome of very confiderable length, their breadth and length in fome degree correfponding to one another; and when longeft they are commonly the broadeft, but not always fo. (See Tab. XXII.) Thefe plates are very different in fize in different parts of the fame mouth, more efpecially in the large Whalebone Whale, whofe upper jaw does not pafs parallel upon the under, but makes an arch, the femidiameter of which is about one-fourth of the length of the jaw. The head in my poffeffion is nineteen feet long, the femidiameter not quite five feet: if this proportion is preferved, thofe Whales which have whalebone fifteen feet long muft be of an immenfe fize.

Thefe plates are placed in feveral rows, encompaffing the outer fkirts of the upper jaw, fimilar to teeth in other animals. They ftand parallel to each other, having one edge towards the circumference of the mouth, the other towards the center or cavity. They are placed near together in the Piked Whale, not being a quarter of an inch afunder where at the greateft diftance, yet differing in this refpect in different parts of the fame mouth; but in the great Whale the diftances are more confiderable.

The outer row is compofed of the longeft plates; and thefe are in proportion to the different diftances between the two jaws, fome being fourteen or fifteen feet long, and twelve or fifteen inches broad; but towards the anterior and pofterior part of the mouth, they are very fhort: they rife for half a foot or more, nearly of equal breadths, and afterwards fhelve off from their inner fide until they come near to a point at the

* From this it muft appear, that the term bone is an improper one.

outer: the exterior of the inner rows are the longeft, correfponding to the termination of the declivity of the outer, and
become fhorter and fhorter till they hardly rife above the gum.

The inner rows are clofer than the outer, and rife almoft
perpendicularly from the gum, being longitudinally ftraight,
and have lefs of the declivity than the outer. The
plates of the outer row laterally are not quite flat, but
make a ferpentine line, more efpecially in the Piked Whale
the outer edge is thicker than the inner. All round the line
made by their outer edges, runs a fmall white bead, which is
formed along with the whalebone, and wears down with it. The
fmaller plates are nearly of an equal thicknefs upon both edges.
In all of them, the termination is in a kind of hair, as if the
plate was fplit into innumerable fmall parts, the exterior being
the longeft and ftrongeft.

The two fides of the mouth compofed of thefe rows meet
nearly in a point at the tip of the jaw, and fpread or recede laterally from each other as they pafs back; and at their pofterior
ends, in the Piked Whale, they make a fweep inwards, and come
very near each other, juft before the opening of the œfophagus. In the Piked Whale there were above three hundred in
the outer rows on each fide of the mouth. Each layer terminates in an oblique furface, which obliquity inclines to the roof
of the mouth, anfwering to the gradual diminution of their
length; fo that the whole furface, compofed of thefe terminations, forms one plane rifing gradually from the roof of the
mouth; from this obliquity of the edge of the outer row, we
may in fome meafure judge of the extent of the whole bafe,
but not exactly, as it makes a hollow curve, which increafes
the bafe.

The

The whole furface refembles the fkin of an animal covered with ftrong hair, under which furface the tongue muft immediately lie, when the mouth is fhut; it is of a light-brown colour in the Piked Whale, and is darker in the large Whale.

In the Piked Whale, when the mouth is fhut, the projecting whalebone remains entirely on the infide of the lower jaw, the two jaws meeting every where along their furface; but how this is effected in the large Whale I do not certainly know, the horizontal plane made by the lower jaw being ftraight, as in the Piked Whale; but the upper jaw being an arch cannot be hid by the lower. I fuppofe, therefore, that a broad upper lip, meeting as low as the lower jaw, covers the whole of the outer edges of the exterior rows.

The whalebone is continually wearing down, and renewing in the fame proportion, except when the animal is growing it is renewed fafter, and in proportion to the growth.

The formation of the whalebone is extremely curious, being in one refpect fimilar to that of the hair, horns, fpurs, &c.; but it has befides another mode of growth and decay, equally fingular.

Thefe plates form upon a thin vafcular fubftance, not immediately adhering to the jaw-bone; but having a more denfe fubftance between, which is alfo vafcular. This fubftance, which may be called the nidus of the whalebone, fends out (the above) thin broad proceffes, anfwering to each plate, on which the plate is formed, as the Cock's fpur or the Bull's horn, on the bony core, or a tooth on its pulp; fo that each plate is necceffarily hollow at its growing end, the firft part of the growth taking place on the infide of this hollow.

Befides

Beſides this mode of growth, which is common to all ſuch ſubſtances, it receives additional layers on the outſide, which are formed upon the above-mentioned vaſcular ſubſtance extended along the ſurface of the jaw. This part alſo forms upon it a ſemi-horny ſubſtance between each plate, which is very white, riſes with the whalebone, and becomes even with the outer edge of the jaw, and the termination of its outer part forms the bead above mentioned. This intermediate ſubſtance fills up the ſpaces between the plates as high as the jaw, acts as abutments to the whalebone, or is ſimilar to the alveolar proceſſes of the teeth, keeping them firm in their places. (See Tab. XXIII.)

As both the whalebone and intermediate ſubſtance are conſtantly growing, and as we muſt ſuppoſe a determined length neceſſary, a regular mode of decay muſt be eſtabliſhed, not depending entirely on chance, or the uſe it is put to.

In its growth, three parts appear to be formed; one from the riſing core, which is the center, a ſecond on the outſide, and a third being the intermediate ſubſtance. Theſe appear to have three ſtages of duration; for that which forms on the core, I believe, makes the hair, and that on the outſide makes principally the plate of whalebone; this, when got a certain length, breaks off, leaving the hair projecting, becoming at the termination very brittle; and the third, or intermediate ſubſtance, by the time it riſes as high as the edge of the ſkin of the jaw, decays and ſoftens away like the old cuticle of the ſole of the foot when ſteeped in water.

The uſe of the whalebone, I ſhould believe, is principally for the retention of the food till ſwallowed; and do ſuppoſe the fiſh they catch are ſmall, when compared with the ſize of the mouth.

5   The

The œfophagus, as in other animals, begins at the fauces, or pofterior part of the mouth; and, although circular at this part, is foon divided into two paffages by the epiglottis paffing acrofs it, as will be defcribed hereafter. Below its attachment to the trachea, it paffes down in the pofterior mediaftinum, at fome diftance from the fpine, to which it is attached by a broad part of the fame membrane, and its anterior furface makes the pofterior part of a cavity behind the pericardium.

Paffing through the diaphragm it enters the ftomach, and is lined with a very thick, foft, and white cuticle, which is continued into the firft cavity of the ftomach.

The inner, or true coat, is white, of a confiderable denfity, and not mufcular; but thrown into large longitudinal folds by the contraction of the mufcular fibres of the œfophagus, which are very ftrong. It is very glandular; for on its inner furface, efpecially near the fauces, orifices of a vaft number of glands are vifible.

The œfophagus is larger in proportion to the bulk of the animal than in the quadruped, although not fo much fo as it ufually is in fifh, which we may fuppofe fwallow their food much in the fame way. In the Piked Whale it was three inches and an half wide.

The ftomach, as in other animals, lies on the left fide of the body, and terminates in the pylorus towards the right.

The duodenum paffes down on the right fide, very much as in the human fubject, excepting that it is more expofed from the colon not croffing it. It lies on the right kidney, and then paffes to the left fide behind the afcending part of the colon and root of the mefentery, comes out on the left fide, and getting on the edge of the mefentery becomes a loofe inteftine,

teſtine, forming the jejunum. In this courſe behind the meſentery it is expoſed, as in moſt quadrupeds, not being covered by it,
as in the human. The jejunum and ilium paſs along the edge of
the meſentery downwards to the lower part of the abdomen.
The ilium near the lower end makes a turn towards the
right ſide, and then mounting upwards, round the edge of
the meſentery, paſſes a little way on the right, as high as
the kidney, and there enters the colon, or cæcum. The cæcum lies on the lower end of the kidney, conſiderably higher
than in the human body, which renders the aſcending part of
the colon ſhort. The cæcum is about ſeven inches long, and
more like that of the Lion or Seal than of any other animal I
know.

The colon paſſes obliquely up the right ſide, a little towards
the middle of the abdomen ; and when as high as the ſtomach,
croſſes to the left, and acquires a broad meſocolon : at this
part it lies upon the left kidney, and in its paſſage down gets
more and more to the middle line of the body. When it has
reached the lower part of the abdomen, it paſſes behind the
uterus, and along with the vagina, in the female ; between the
two teſticles, and behind the bladder and root of the penis, in
the male, bending down to open on what is called the belly
of the animal ; and in its whole courſe it is gently convoluted.
In thoſe which have no cæcum, and therefore can hardly be
ſaid to have a colon, the inteſtine before its termination in the
rectum makes the ſame kind of ſweep round the other inteſtines, as the colon does where there is a cæcum.

The inteſtines are not large for the ſize of the animal, not being larger in thoſe of eighteen or twenty-four feet long than in
the Horſe, the colon not much more capacious than the jejunum
and ilium, and very ſhort ; a circumſtance common to carni-

vorous

vorous animals. In the Piked Whale, the length from the
ftomach to the cæcum is 28 yards and an half, length of cæcum
feven inches, of the colon to the anus two yards and three
quarters. The fmall inteftines are juft five times the length
of the animal, the colon with the cæcum a little more than
one-half the length.

Thofe parts that refpect the nourifhment of this tribe do not
all fo exactly correfpond as in land animals; for in thefe
one in fome degree leads to the other. Thus the teeth in the
ruminating tribe point out the kind of ftomach, cæcum, and
colon; while in others, as the Horfe, Hare, Lion, &c. the
appearances of the teeth only give us the kind of colon and
cæcum; but in this tribe, whether teeth or no teeth, the ftomachs do not vary much, nor does the circumftance of cæcum
feem to depend on either teeth or ftomach. The circumftances
by which, from the form of one part we judge what others
are, fail us here; but this may arife from not knowing all the
circumftances. The ftomach, in all that I have examined,
confifts of feveral bags, continued from the firft on the left
towards the right, where the laft terminates in duodenum.
The number is not the fame in all; for in the Porpoife, Grampus, and Piked Whale, there are five; in the Bottle-nofe
feven. Their fize refpecting one another differs very confiderably; fo that the largeft in one fpecies may in another be only
the fecond. The two firft in the Porpoife, Bottle-nofe, and
Piked Whale, are by much the largeft; the others are fmaller,
although irregularly fo.

The firft ftomach has, I believe, in all very much the fhape
of an egg, with the fmall end downwards. It is lined every
where with a continuation of the cuticle from the œfophagus. In
the Porpoife the œfophagus enters the fuperior end of the fto-

mach. In the Piked Whale its entrance is a little way on the poſterior part of the upper end, and is oblique.

The ſecond ſtomach in the Piked Whale is very large, and rather longer than the firſt. It is of the ſhape of the Italic *S*, paſſing out from the upper end of the firſt on its right ſide, by nearly as large a beginning as the body of the bag. In the Porpoiſe it by no means bears the ſame proportion to the firſt, and opens by a narrower orifice; then paſſing down along the right ſide of the firſt ſtomach, it bends a little outwards at the lower end, and terminates in the third. Where this ſecond ſtomach begins, the cuticle of the firſt ends. The whole of the inſide of this ſtomach is thrown into unequal rugæ, appearing like a large irregular honeycomb. In the Piked Whale the rugæ are longitudinal, and in many places very deep, ſome of them being united by croſs bands; and in the Porpoiſe the folds are very thick, maſſy, and indented into one another. This ſtomach opens into the third by a round contracted orifice, which does not ſeem to be valvular.

The third ſtomach is by much the ſmalleſt, and appears to be only a paſſage between the ſecond and fourth. It has no peculiar ſtructure on the inſide, but terminates in the fourth by nearly as large an opening as its beginning. In the Porpoiſe it is not above one, and in the Bottle-noſe about five inches long.

The fourth ſtomach is of a conſiderable ſize; but a good deal leſs than either the firſt or ſecond. In the Piked Whale it is not round, but ſeems flattened between the ſecond and fifth. In the Porpoiſe it is long, paſſing in a ſerpentine courſe almoſt like an inteſtine. The internal ſurface is regular, but villous, and opens on its right ſide into the fifth, by a round opening ſmaller than the entrance from the third.

The

The fifth ftomach is in the Piked Whale round, and in the Porpoife oval; it is fmall, and terminates in the pylorus, which has little of a valvular appearance. Its coats are thinner than thofe of the fourth, having an even inner furface, which is commonly tinged with bile.

The Piked Whale and, I believe, the large Whalebone Whale, have a cæcum; but it is wanting in the Porpoife, Grampus, and Bottle-nofe Whale.

The ftructure of the inner furface of the inteftine is in fome very fingular, and different from that of the others.

The inner furface of the duodenum in the Piked Whale is thrown into longitudinal rugæ, or valves, which are at fome diftance from each other, and thefe receive lateral folds. The duodenum in the Bottle-nofe fwells out into a large cavity, and might almoft be reckoned an eighth ftomach; but as the gall ducts enter it I fhall call it duodenum.

The inner coat of the jejunum, and ilium, appears in irregular folds, which may vary according as the mufcular coat of the inteftine acts: yet I do not believe, that their form depends intirely on that circumftance, as they run longitudinally, and take a ferpentine courfe when the gut is fhortened by the contraction of the longitudinal mufcular fibres. The inteftinal canal of the Porpoife has feveral longitudinal folds of the inner coat paffing along it, through the whole of its length. In the Bottle-nofe the inner coat, through nearly the whole track of the inteftine, is thrown into large cells, and thefe again fubdivided into fmaller; the axis of which cells is not perpendicular to a tranfverfe fection of the inteftine, but oblique, forming pouches with the mouths downwards, and acting almoft like valves, when any thing is attempted to be paffed in a contrary direction: they begin faintly in the duodenum, before it makes

its

its quick turn, and terminate near the anus. The colon and rectum have the rugæ very flat, which ſeems to depend intirely on the contraction of the gut.

The rectum near the anus appears, for four or five inches, much contracted, is glandular, covered by a ſoft cuticle, and the anus ſmall.

I never found any air in the inteſtines of this tribe; nor indeed in any of the aquatic animals.

The meſenteric artery anaſtomoſes by large branches.

There is a conſiderable degree of uniformity in the liver of this tribe of animals. In ſhape it nearly reſembles the human, but is not ſo thick at the baſe, nor ſo ſharp at the lower edge, and is probably not ſo firm in its texture. The right lobe is the largeſt and thickeſt, its falciform ligament broad, and there is a large fiſſure between the two lobes, in which the round ligament paſſes. The liver towards the left is very much attached to the ſtomach, the little epiploon being a thick ſubſtance. There is no gall-bladder; the hepatic duct is large, and enters the duodenum about ſeven inches beyond the pylorus.

The pancreas is a very long, flat body, having its left end attached to the right ſide of the firſt cavity of the ſtomach: it paſſes acroſs the ſpine at the root of the meſentery, and near to the pylorus joins the hollow curve of the duodenum, along which it is continued, and adheres to that inteſtine, its duct entering that of the liver near the termination in the gut.

Although this tribe cannot be ſaid to ruminate, yet in the number of ſtomachs they come neareſt to that order; but here I ſuſpect that the order of digeſtion is in ſome degree inverted. In both the ruminants, and this tribe, I think it muſt be allowed that the firſt ſtomach is a reſervoir. In the ruminants the preciſe uſe of the ſecond and third ſtomachs is perhaps not known;

but

but digeftion is certainly carried on in the fourth; while in this tribe, I imagine, digeftion is performed in the fecond, and the ufe of the third and fourth is not exactly afcertained.

The cæcum and colon do not aflift in pointing out the nature of the food and mode of digeftion in this tribe. The Porpoife which has teeth, and four cavities to the ftomach, has no cæcum, fimilar to fome land animals, as the Bear, Badger, Racoon, Ferret, Polecat, &c.; neither has the Bottle-nofe a cæcum which has only two fmall teeth in the lower jaw; and the Piked Whale, which has no teeth, has a cæcum, almoft exactly like the Lion, which has teeth and a very different kind of ftomach.

The food of the whole of this tribe, I believe, is fifh; probably each may have a particular kind, of which it is fondeft, yet does not refufe a variety. In the ftomach of the large Bottle-nofe, I found the beaks of fome hundreds of Cuttle-fifh. In the Grampus I found the tail of a Porpoife; fo that they eat their own genus. In the ftomach of the Piked Whale, I found the bones of different fifh, but particularly thofe of the Dog-fifh. From the fize of the œfophagus we may conclude, that they do not fwallow fifh fo large in proportion to their fize as many fifh do, that we have reafon to believe take their food in the fame way: for fifh often attempt to fwallow what is larger than their ftomachs can at one time contain, and part remains in the œfophagus till the reft is digefted.

The epiploon on the whole is a thin membrane; on the right fide it is rather a thin net-work, though on the left it is a complete membrane, and near to the ftomach of the fame fide becomes of a confiderable thicknefs, efpecially between the two firft bags of the ftomach. It has little or no fat, except what

what flightly covers the veffels in particular parts. It is attached
forwards, all along to the lower part of the different bags confti-
tuting the ftomach, and on the right to the root of the mefentery, between the ftomach and tranfverfe arch of the colon, firft
behind to the tranfverfe arch of the colon and root of the mefentery, then to the pofterior furface of the left or firft bag of
the ftomach, behind the anterior attachment. In fome of this
tribe there is the ufual paffage behind the veffels going to the
liver, common to all quadrupeds I am acquainted with; but in
others, as the fmall Bottle-nofe, there is no fuch paffage,
which by the cavity behind the ftomach in the epiploon of this
animal becomes a circumfcribed cavity.

The fpleen is involved in the epiploon, and is very fmall for
the fize of the animal. There are in fome, as the Porpoife,
one or two fmall ones, about the fize of a nutmeg, often fmaller, placed in the epiploon behind the other. Thefe are fome-
times met with likewife in the human body.

The kidnies in the whole of this tribe of animals are con-
glomerated, being made up of fmaller parts, which are only
connected by cellular membrane, blood-veffels, and ducts, or
infundibula; but not partially connected by continuity of fubftance, as in the human body, the Ox, &c.: every portion is of
a conical figure, whofe apex is placed towards the center of
the kidney, the bafe making the external furface; and each is
compofed of a cortical and tubular fubftance, the tubular terminating in the apex, which apex makes the mamilla. Each
mamilla has an infundibulum, which is long, and at its beginning wide, embracing the bafe of the mamilla, and becoming fmaller. Thefe infundibula unite at laft, and form the
ureter. The whole kidney is an oblong flat body, broader
and thicker at the upper end than the lower, and has the

appearance

appearance of being made up of different parts placed close together, almost like the pavement of a street.

The ureter comes out at the lower end, and passes along to the bladder, which it enters very near the urethra.

The bladder is oblong, and small for the size of the animal. In the female the urethra passes along to the external sulcus or vulva, and opens just under the clitoris, much as in the human subject.

Whether being inhabitants of the water makes such a construction of kidney necessary I cannot say; yet one must suppose it to have some connection with such situation, since we find it almost uniformly take place in animals inhabiting the water, whether wholly, as this tribe, or occasionally, as the Manatee, Seal, and White Bear: there is, however, the same structure in the Black Bear, which, I believe, never inhabits the water. This, perhaps, should be considered in another light, as nature keeping up to a certain uniformity in the structure of similar animals; for the Black Bear in construction of parts is, in every other respect as well as this, like the White Bear.

The capsulæ renales are small for the size of the animal, when compared to the human, as indeed they are in most animals. They are flat, and of an oval figure; the right lies on the lower and posterior part of the diaphragm somewhat higher than the kidney; the left is situated lower down, by the side of the aorta, between it and the left kidney. They are composed of two substances; the external having the direction of its fibres or parts towards the center; the internal seeming more uniform, and not having so much of the fibrous appearance.

The blood of animals of this order is, I believe, similar to that of quadrupeds; but I have an idea, that the red globules

are.

are in larger proportion. I will not pretend to determine how far this may aſſiſt in keeping up the animal heat; but as theſe animals may be ſaid to live in a very cold climate or atmoſphere, and ſuch as readily carries off heat from the body, they may want ſome help of this kind.

It is certain that the quantity of blood in this tribe and in the Seal is comparatively larger than in the quadruped, and therefore probably amounts to more than that of any other known animal.

This tribe differs from fiſh in having the red blood carried to the extreme parts of the body, ſimilar to the quadruped.

The cavity of the thorax is compoſed of nearly the ſame parts as in the quadruped; but there appears to be ſome difference, and the varieties in the different genera are greater.

The general cavity is divided into two, as in the quadruped, by the heart and mediaſtinum.

The heart in this tribe, and in the Seal, is probably larger in proportion to their ſize than in the quadruped, as alſo the blood-veſſels, more eſpecially the veins.

The heart is incloſed in its pericardium, which is attached by a broad ſurface to the diaphragm, as in the human body. It is compoſed of four cavities *, two auricles, and two ventricles : it is more flat than in the quadruped, and adapted to the ſhape of the cheſt. The auricles have more faſciculæ, and theſe paſs more acroſs the cavity from ſide to ſide than in many other animals; beſides, being very muſcular, they are very elaſtic,

---

* As the circulation is a permanent part of the conſtitution reſpecting the claſs to which the animal belongs, and as the kind of heart correſponds with the circulation, theſe ſhould be conſidered in the claſſing of animals. Thus we have animals whoſe hearts have only one cavity, others with two, three, and four cavities.

7   for

for being ftretched they contract again very confiderably. There is nothing uncommon or particular in the ftructure of the ventricles, in the valves of the ventricles, or in that of the arteries.

The general ftructure of the arteries refembles that of other animals; and where parts are nearly fimilar, the diftribution is likewife fimilar. The aorta forms its ufual curve, and fends off the carotid and fubclavian arteries.

Animals of this tribe, as has been obferved, have a greater proportion of blood than any other known, and there are many arteries apparently intended as refervoirs, where a larger quantity of arterial blood feemed to be required in a part, and vafcularity could not be the only object. Thus we find, that the intercoftal arteries divide into a vaft number of branches, which run in a ferpentine courfe between the pleura, ribs, and their mufcles, making a thick fubftance fomewhat fimilar to that formed by the fpermatick artery in the Bull. Thofe veffels, every where lining the fides of the thorax, pafs in between the ribs near their articulation, and alfo behind the ligamentous attachment of the ribs, and anaftomofe with each other. The medulla fpinalis is furrounded with a net-work of arteries in the fame manner, more efpecially where it comes out from the brain, where a thick fubftance is formed by their ramifications and convolutions; and thefe veffels moft probably anaftomofe with thofe of the thorax.

The fubclavian artery in the Piked Whale, before it paffes over the firft rib, fends down into the cheft arteries which affift in forming the plexus on the infide of the ribs; I am not certain but the internal mammary arteries contribute to form the anterior part of this plexus. The motion of the blood in fuch muft be very flow; the ufe of which we do not readily fee. The defcending aorta fends off the intercoftals, which are very large, and give branches to this plexus; and when it has reached the abdomen, it fends off, as in the quadruped, the different

N n n branches

branches to the vifcera, and the lumbar arteries, which are likewife very large for the fupply of that vaft mafs of mufcles
which moves the tail.

In our examination of particular parts, the fize of which is
generally regulated by that of the whole animal, if we have only
been accuftomed to fee them in thofe which are fmall or middle-
fized, we behold them with aftonifhment in animals fo far
exceeding the common bulk as the Whale. Thus the heart
and aorta of the Spermaceti Whale appeared prodigious, being
too large to be contained in a wide tub, the aorta mea-
furing a foot in diameter. When we confider thefe as applied
to the circulation, and figure to ourfelves, that probably ten
or fifteen gallons of blood are thrown out at one ftroke, and
moved with an immenfe velocity through a tube of a foot diameter, the whole idea fills the mind with wonder.

The veins, I believe, have nothing particular in their ftructure, excepting in parts requiring a peculiarity, as in the folds
of the fkin on the breaft in the Piked Whale, where their elafticity was to be increafed.

Of the Larynx.

The larynx in moft animals living on land is a compound
organ, adapted both for refpiration, deglutition, and found, which
laft is produced in the actions of refpiration ; but in this tribe
the larynx, I fuppofe, is only adapted to refpiration, as we do
know that they have any mode of producing found.

It is compofed of os hyoides, thyroid, cricoid, and two arytenoid cartilages, with the epiglottis. It varies very much in
ftructure and fize, when compared in animals of different
genera. Thefe cartilages were much fmaller in the Bottle-
nofe of twenty-four feet long, than in the Piked Whale of
feventeen feet, while the os hyoides was much larger.

In

In the Bottle-nose, the os hyoides is compofed of three bones, befides two whofe ends are attached to it, being placed above the os hyoides, making five in all. In the Porpoife, Piked Whale, &c. it is but one bone, flightly bent, having a broad thin procefs paffing up, which is a little forked: it has no attachment to the head by means of other bones, as in many quadrupeds.

The thyroid cartilage in the Piked Whale is broad from fide to fide, but not from the upper to the lower part: it has two lateral proceffes, which are long, and pafs down the outfide of the cricoid, near to its lower end, and are joined to it much as in the human fubject. Thefe differ in fhape in different animals of this tribe.

The cricoid cartilage is broad and flat, making the pofterior and lateral part of the larynx, and is much deeper behind, and laterally, than before. It is extremely thick and ftrong, flattened on the pofterior furface, and hollowed from the upper edge to the lower. It terminates by a thick edge on the pofterior part above, but irregularly at the lower edge, in the cartilages of the larynx.

The two arytenoid cartilages are extremely projecting, and united to each other till near their ends; are articulated on the upper edge of the cricoid, but fend down a procefs, which paffes on the infide of the cricoid, being attached to a bag in the Piked Whale, which is formed below the thyroid and before the cricoid cartilages; they crofs the cavity of the larynx obliquely, making the paffage, at the upper part, a groove between them: the cavity at this place fwells out laterally, but is very narrow between the anterior and pofterior furfaces. The paffage above between the arytenoid and thyroid cartilages is wide from fide to fide, and is continued down on the outfide of the proceffes of

the

the arytenoid cartilage, as well as between them, ending below the thyroid, which is folliculated on its inner furface on the fore part of the cricoid cartilage.

The epiglottis makes a third part of the paffage, and compleats the glottis by forming it into a canal, in feveral of this tribe ; but in the Piked Whale it was not attached to the two arytenoid cartilages, but only in contact, or inclofing them at their bafe, fo as to make them form a complete canal.

I could not obferve any thing like a thyroid gland.

From the glottis and epiglottis being fo connected as to make but one canal, and from the thyroid and cricoid cartilages being fo flattened in fome between the anterior and pofterior furface, the paffage through thefe parts is very fmall or con-tracted ; but the trachea fwells out again into a very confidera-ble fize. Its larger branches are in proportion to the trunk, and enter the lungs at the upper end along with the blood-veffels.

Of the Lungs.

The lungs are two oblong bodies, one on each fide of the cheft, and are not divided into fmaller lobes, as in the human fubject. They are of confiderable length, but not fo deep between the fore and back part, as in the quadruped, from the heart being broad, flat, and of itfelf filling up the fore part of the cheft. They pafs further down on the back part than in the quadruped, by which their fize is increafed, and rife higher up in the cheft than the entrance of the veffels, coming to a point at the upper end. From the entrance of the veffels they are connected downwards, along their whole inner edge, by a ftrong attachment (in which there are in fome lymphatic glands) to

3   the

the posterior mediastinum. The lungs are extremely elastic in their substance, even so much so as to squeeze out any air that may be thrown into them, and to become almost at once a solid mass, having a good deal the appearance, confistence, and feel of an ox's spleen. The branches of the bronchiæ which ramify into the lungs have not the cartilages flat, but rather rounded; a conftruction which admits of greater motion between each.

The pulmonary cells are fmaller than in quadrupeds, which may make lefs air neceffary, and they communicate with each other, which thofe of the quadruped do not; for by blowing into one branch of the trachea, not only the part to which it immediately goes, but the whole lungs are filled.

As the ribs in this tribe do not completely make the cavity of the thorax, the diaphragm has not the fame attachments as in the quadruped, but is connected forwards to the abdominal mufcles, which are very ftrong, being a mixture of mufcular and tendinous fibres.

The pofition of the diaphragm is lefs tranfverfe than in the quadruped, paffing more obliquely backwards, and coming very low on the fpine, and higher up before; which makes the cheft longeft in the direction of the animal at the back, and gives room for the lungs to be continued along the fpine.

The parts immediately concerned in infpiration are extremely ftrong; the diaphragm remarkably fo. The reafon of this muft at once appear; it neceffarily requiring great force to expand in a denfe medium like water, efpecially too when the vacuity is to be filled with one which is rarer, and is to water a fpecies of vacuum, the preffure being much greater on the external furface than the counter-preffure from within. But expiration on the other hand muft be much more eafily performed;

formed; the natural elaſticity of the parts themſelves, with the preſſure of the water on the external ſurface of the body, being greater than the reſiſtance of the air from within, will both tend to produce expiration without any immediate action of muſcles.

The diaphragm, in theſe animals, appears to be the principal agent in inſpiration; and the cavity of the thorax not being intirely ſurrounded by bony parts, is of courſe leſs eaſily expanded, and the apparatus for its expanſion in all directions, as in the quadruped, does not exiſt here.

The Blow-hole, or Paſſage for the Air.

As the noſe in every animal that breathes air is a common paſſage for the air, and is alſo the organ of ſmelling; I ſhall deſcribe it in this tribe as inſtrumental to both theſe purpoſes.

There is a variety in ſome ſpecies of this animal which is, I believe, peculiar to this order; that is, the want of the ſenſe of ſmelling; none of thoſe which I have yet examined having that ſenſe, except the two kinds of Whalebone Whale: ſuch of courſe have neither the olfactory nerves, nor the organ; therefore, in them, the noſtrils are intended merely for reſpiration; but others have the organ placed in this paſſage as in other animals.

The membranous portion of the poſterior noſtrils is one canal; but when in the bony part, in moſt of them, it is divided into two; the Spermaceti Whale, however, is an exception. In thoſe which have it divided, it is in ſome continued double through the anterior ſoft parts, opening by two orifices, as in the Piked Whale; but in others, it unites again in the membranous part, making externally only one orifice, as in the

7   Porpoiſe,

Porpoife, Grampus, and Bottle-nofe. At its beginning in the fauces, it is a roundifh hole, furrounded by a ftrong fphincter mufcle, for grafping the epiglottis; beyond this, the canal becomes larger, and opens into the two paffages in the bones of the head. This part is very glandular, being full of follicles, whofe ducts ramify in the furrounding fubftance, which appears fatty and mufcular like the root of the tongue, and thefe ramifications communicate with one another, and contain a vifcid flime.

In the Spermaceti Whale, which has a fingle canal, it is thrown a little to the left fide. After thefe canals emerge from the bones near the external opening, they become irregular, and have feveral fulci paffing out laterally, of irregular forms, with correfponding eminences. The ftructure of thefe eminences is mufcular and fatty, but lefs mufcular than the tongue of a quadruped.

In the Porpoife there are two fulci on each fide; two large and two fmall, with correfponding eminences of different fhapes, the large ones being thrown into folds. The Spermaceti Whale has the leaft of this ftructure; the external opening in it comes farther forwards towards the anterior part of the head, and is confequently longer than in others of this order. Near to its opening externally, it forms a large fulcus, and on each fide of this canal is a cartilage, which runs nearly its whole length. In all that I have examined, this canal, forwards from the bones, is intirely lined with a thick cuticle of a dark colour.

In thofe which have only one external opening, it is tranfverfe, as in the Porpoife, Grampus, Bottle-nofe and Spermaceti Whale, &c.; where double, they are longitudinal, as in the Piked Whale, and the large Whalebone Whale. Thefe
openings

openings form a paffage for the air in refpiration to and from
the lungs; for it would be impoffible for thefe animals to
breathe air through the mouth; indeed, I believe, the human
fpecies alone breathe by the mouth, and in them it is moftly
from habit; for in quadrupeds the epiglottis conducts the
air into the nofe.

In the whole of this tribe, the fituation of the opening on
the upper furface of the head is well adapted for this purpofe,
being the firft part that comes to the furface of the water in
the natural progreffive motion of the animal; therefore it is to
be confidered principally as a refpiratory organ, and where it
contains the organ of fmell, that is only fecondary.

As the animals of this order do not live in the medium which
they infpire, the organs conducting the air to the lungs are in
fome fort particularly conftructed, that the water in which
they live may not interfere with the air they breathe.

The projecting glottis, which has been defcribed, paffes into
the pofterior noftrils, by which means it croffes the fauces, dividing them into two paffages. The enlargement at the termination of the glottis, obferved in fome of them, would feem
to be intended to prevent its retraction; but, as it feems confined to the Porpoife and Grampus, it may, perhaps, in them
anfwer fome other purpofe.

The beginning of the pofterior noftrils, which anfwers to the
palatum molle in the quadruped, having a fphincter, the glottis is grafped by it, which renders its fituation ftill more fecure,
and the paffages through the head, acrofs the fauces and along the
trachea, are rendered one continued canal; this union of glottis
and epiglottis with the pofterior noftril, making only a kind of
joint, admits of motion, and of dilatation and contraction of the
fauces, in deglutition, from the epiglottis moving more in or
out of the pofterior noftril.

This

This conftruction of parts anfwers a purpofe fimilar to that
of the epiglottis in the quadruped ; it may be confidered as the
epiglottis and the arytenoid cartilages joining, to make a
tubular or cylindrical epiglottis, inftead of a valvular one.

The reafons why there fhould be fo peculiar a conftruction of
parts do not at firft appear ; but we certainly fee by it an
abfolute guard placed upon the lungs, that no water fhould get
into them.

This tribe being without the projecting tongue of the quadruped, and wanting its extenfive motion, and the power
of fucking things into the mouth, may probably require
the conftruction between the air and lungs to be more perfect ;
but how far it is fo, I will not pretend to fay.

The fize of the Brain differs much in different genera of this
tribe, and likewife in the proportion it bears to the bulk of
the animal. In the Porpoife, I believe, it is largeft, and
perhaps in that refpect comes neareft to the human.

The fize of the cerebellum in proportion to that of the cerebrum is fmaller in the human fubject than in any animal with
which I am acquainted. In many quadrupeds, as the Horfe,
Cow, &c. the difproportion in fize between cerebellum and
cerebrum is not great, and in this tribe it is ftill lefs ; yet not
fo fmall as in the bird, &c.

The whole brain in this tribe is compact, the anterior
part of the cerebrum not projecting fo far forwards as in either
the quadruped or in the human fubject ; neither is the medulla
oblongata fo prominent, but flat, lying in a kind of hollow
made by the two lobes of the cerebellum.

The brain is compofed of cortical and medullary fubftances,
very diftinctly marked ; the cortical being, in colour, like the
tubular fubftance of a kidney ; the medullary, very white.

Thefe fubftances are nearly in the fame proportion as in the human brain. The two lateral ventricles are large, and in thofe that have olfactory nerves are not continued into them as in many quadrupeds ; nor do they wind fo much outwards as in the human fubject, but pafs clofe round the pofterior ends of the thalami nervorum opticorum. The thalami them-felves are large; the corpora ftriata fmall; the crura of the fornix arc continued along the windings of the ventriclès, much as in the human fubject. The plexus choroides is attached to a ftrong membrane, which covers the thalami nervorum opti-corum, and paffes through the whole courfe of the ventricle, much as in the human fubject.

The fubftance of the brain is more vifibly fibrous than I ever faw it in any other animal, the fibres paffing from the ven-tricles as from a center to the circumference, which fibrous texture is alfo continued through the cortical fubftance. The whole brain in the Piked Whale weighed four pounds ten ounces.

The nerves going out from the brain, I believe, are fimilar to thofe of the quadruped, except in the want of the olfactory nerves in the genus of the Porpoife.

The medulla fpinalis is much fmaller in proportion to the fize of the body than in the human fpecies, but ftill bears fome proportion to the quantity of brain; for in the Porpoife, where the brain is largeft, the medulla fpinalis is largeft ; yet this did not hold good in the Spermaceti Whale, the fize of the me-dulla fpinalis appearing to be proportionally larger than the brain, which was fmall when compared to the fize of the animal. It has a cortical part in the center, and terminates about the twenty-fifth vertebra, beyond which is the cauda equina, the dura mater going no lower. The nerves which go

2   off

off from the medulla fpinalis are more uniform in fize than in the quadruped, there being no fuch inequality of parts, nor any extremities to be fupplied, except the fins.

The medulla fpinalis is more fibrous in its ftructure than in other animals; and when an attempt is made to break it longitudinally, it tears with a fibrous appearance, but tranfverfely it breaks irregularly.

The dura mater lines the fkull, and forms in fome the three proceffes anfwerable to the divifions of the brain, as in the human fubject; but in others, this is bone. Where it covers the medulla fpinalis, it differs from all the quadrupeds I am acquainted with, inclofing the medulla clofely, and the nerves immediately paffing out through it at the lower part, as they do at the upper, fo that the cauda equina, as it forms, is on the outfide of the dura mater.

As the Organs of Senfe are varioufly formed in different animals, fitted for the various modes of impreffion; and as the modes are either increafed or varied, according to circumftances which make no part of the fenfe itfelf, but which are neceffary for the œconomy of the animal, we find the fenfes in this tribe varied in their conftruction, and in fome a fenfe is even wholly wanting.

The organs of fenfe, which appear to be adapted to every mode of life, are thofe of touch and tafte; but thofe of fmell, fight, and hearing, probably require to be varied according to circumftances. Thus fmell may be increafed by a mode of impregnation, hearing by the vibration of different mediums, and fight by the different powers of refraction of different mediums; therefore, as animals are intended by nature to be differently circumftanced, fo are the fenfes formed.

Of

Of the Senfe of Touch.

The cutis in this tribe appears, in general, particularly well calculated for fenfation ; the whole furface being covered with villi, which are fo many veffels, and we muft fuppofe, nerves. Whether this ftructure is only neceffary for acute fenfation, or whether it is neceffary for common fenfation, where the cuticle is thick, and confifting of many layers, I do not know. We may obferve, that where it is neceffary the fenfe of touch fhould be accurate, the villi are ufually thick and long, which probably is neceffary, becaufe in moft parts of the body, where the more acute fenfations of touch are required, fuch parts are covered by a thick cuticle. Of this the ends of our fingers, toes, and the foot of the hoofed animals, are remarkable examples.

Whether this fenfe is more acute in water, I am not certain, but fhould imagine it is.

Of the Senfe of Tafte.

The tongue, which is the organ of tafte, is alfo endowed with the fenfe of touch. It is likewife to be confidered, in the greateft number of animals, as an inftrument for mechanical purpofes ; but probably lefs fo in this tribe than any other. However, even in thefe, it muft have been formed with this view, fince, merely as an organ of tafte, it would only have required furface, yet is a projecting body endowed with motion. In fome, it is better adapted for motion than in others ; and I fhould fuppofe this to be requifite, on account of the difference in the mode of catching the food, and in the act of fwallowing. It is moft projecting in thofe with teeth, probably for the better

conducting

conducting the food, step by step, to the œsophagus; whereas, it does not seem so necessary to have such management of the tongue in those which have no teeth, and catch their food by merely opening the mouth, and swimming upon it, or by having their prey carried in by the water. In the Porpoise and Grampus it is firm in texture, composed of muscle and fat, being pointed and serrated on its edges, like that of the Hog.

In the Spermaceti Whale the tongue was almost like a feather-bed. In the Piked Whale it was but gently raised, hardly having any lateral edges, and its tip projecting but little, yet, like every other tongue, composed of muscle and fat. The extent between the two jaw bones in this Whale was very considerable, taking in the whole width of the head or upper jaw, and of course including the whalebone. This extent of surface, between jaw and jaw, having but little projection of tongue, is almost flat from side to side, is extremely elastic when contracted, and throws the inner membrane into a vast number of very small folds, that run parallel to one another, but which are again thrown into a close serpentine course by the elasticity of the part in a contrary direction. From the tongue being capable of but little motion, there is only a small mass of muscle required; and from the thinness of the jaw bones, the distance between the lower surface of the mouth and external surface of the skin is but small; and this skin being ribbed, and very elastic, is capable of considerable distention, by which the cavity of the mouth can be enlarged.

The tongue of the large Whalebone Whale, I should suppose, rose in the mouth considerably; the two jaws at the middle being kept at such a distance on account of the whalebone, so that the space between, when the mouth is shut, must be filled up by the tongue.

6

Of

Of the Senſe of Smelling.

In this tribe of animals there is ſomething very remarkable in what relates to the ſenſe of ſmelling; nor have I been able to diſcover the particular mode by which it is performed.

When we conſider theſe animals as quadrupeds, and only conſtructed differently in external form for progreſſive motion through water, we muſt ſee that it was neceſſary that all the ſenſes ſhould correſpond with this medium: we muſt therefore be at a loſs to conceive how they ſmell, ſince we may obſerve, that the organ for ſmelling water, as in fiſh, is very different from that formed to ſmell air; and as we muſt ſuppoſe this tribe are only to ſmell water, being the medium in which ſuch odoriferous particles can be diffuſed, we ſhould expect their organ to be ſimilar to that of fiſh; but in that caſe nature would have been obliged to have attached the noſe of a fiſh to an animal conſtructed like a quadruped; and it is contrary to the laws which are eſtabliſhed in the animal creation to mix parts of different animals together.

In many of this tribe there is no organ of ſmell at all; and in thoſe which have ſuch an organ, it is not that of a fiſh, therefore probably not calculated to ſmell water. It becomes difficult, therefore, to account for the manner in which ſuch animals ſmell the water; and why the others ſhould not have had ſuch an organ *,

---

* Is the mode of ſmelling in fiſh ſimilar to taſting in other animals? Or is the air contained in the water impregnated with the odoriferous parts, and this air the fiſh ſmells? If ſo, it is ſomewhat ſimilar to the breathing of fiſh, it not being the water which produces the effect there, but the air contained in it. This I proved by experiments, and is mentioned by Dr. PRIESTLEY.

which,

which, I believe, is peculiar to the large and small Whale-bone Whales.

Although it is not the external air which they inspire that produces smell, I believe it is the air retained in the nostril out of the current of respiration, which by being impregnated with the odoriferous particles contained in the water during the act of blowing, is applied to the organ of smell. It might be supposed, that they could smell the air on the surface of the water by every inspiration, as animals do on land; and pro-bably they may: but this will not give them the power to smell the odoriferous particles of their prey in the water at any depth; and as their organ is not fitted to be affected by the ap-plication of water, and as they cannot suck water into the nostril, without the danger of its passing into the lungs, it cannot be by its application to this organ that they are enabled to smell.

Some have the power of throwing the water from the mouth through the nostril, and with such force as to raise it thirty feet high: this must answer some important purpose, although not immediately evident to us.

As the organ appears to be formed to smell air only, and as I conceive the smelling of the external air could not be of use as a sense. I therefore believe, that they do not smell in inspi-ration; yet let us consider how they may be supposed to smell the odoriferous particles of the water.

The organ of smell is out of the direct road of the current of air in inspiration; it is also out of the current of water when they spout; may we not suppose then, that this sinus contains air, and as the water passes in the act of throwing it out, that it impregnates this reservoir of air, which imme-diately affects the sense of smell. This operation is probably performed in the time of expiration, because it is said that this

water

water is ſometimes very offenſive; but all this I only give as conjecture.

If the above ſolution is juſt, then only thoſe which have the organ of ſmell can ſpout, a fact worthy of enquiry.

The organ of ſmell would appear to be leſs neceſſary in theſe animals than in thoſe which live in air, ſince ſome are wholly deprived of it; and the organ in thoſe which have it is extremely ſmall, when compared with that of other animals, as well as the nerve which is to receive the impreſſion, as was obſerved above.

Of the Senſe of Hearing.

The ear is conſtructed much upon the ſame principle as in the quadruped; but as it differs in ſeveral reſpects, which it is neceſſary to particulariſe, to convey a perfect idea of it the whole ſhould be deſcribed. As this would exceed the limits of this Paper, I ſhall content myſelf with a general deſcription, taking notice of thoſe material points in which it differs from that of the quadruped.

This organ conſiſts of the ſame parts as in the quadruped; an external opening, with a membrana tympani, an Euſtachian tube, a tympanum with its proceſſes, and the ſmall bones. There is no external projection forming a funnel, but merely an external opening. We can eaſily aſſign a reaſon why there ſhould be no projecting ear, as it would interfere with progreſſive motion; but the reaſon why it is not formed as in birds, is not ſo evident; whether the percuſſions of water could be collected into one point as air, I cannot ſay. The tympanum is conſtructed with irregularities, ſo much like thoſe of an external ear, that I could ſuppoſe it to have a ſimilar effect.

The

The external opening begins by a fmall hole, fcarcely percep-
tible, fituated on the fide of the head a little behind the eye. It is
much longer than in other animals, in confequence of the fize of
the head being fo much increafed beyond the cavity that contains
the brain. It paffes in a ferpentine courfe, at firft horizontally,
then downwards, and afterwards horizontally again, to the membrana tympani, where it terminates. In its whole length it is compofed of different cartilages, which are irregular and united together by cellular membrane, fo as to admit of motion, and probably
of lengthening or fhortening, as the animal is more or lefs fat.

The bony part of the organ is not fo much inclofed in the
bones of the fkull as in the quadruped, confifting commonly
of a diftinct bone or bones, clofely attached to the fkull, but
in general readily to be feparated from it; yet in fome it fends
off, from the pofterior part, proceffes which unite with the fkull.
It varies in its fhape, and is compofed of the immediate organ
and the tympanum.

The immediate organ is, in point of fituation to that of the
tympanum, fuperior and internal, as in the quadruped. The
tympanum is open at the anterior end, where the Euftachian
tube begins.

The Euftachian tube opens on the outfide of the upper part
of the fauces: in fome higher in the nofe than others;
higheft, I believe, in the Porpoife. From the cavity of the
tympanum, where it is rather largeft, it paffes forwards and
inwards, and near its termination appears very much fafcicu-
lated, as if glandular.

The Euftachian tube and tympanum communicate with feveral finufes, which paffing in various directions furround the
bone of the ear. Some of thefe are cellular, fimilar to the
cells of the maftoid procefs in the human fubject, although
not bony. There is a portion of this cellular ftructure of a

particular kind, being white, ligamentous, and each part rather rounded than having flat fides *. One of the finufes paffing out of the tympanum clofe to the membrana tympani, goes a little way in the fame direction, and communicates with a number of cells.

The whole function of the Euftachian tube is perhaps not known; but it is evidently a duct from the cavity of the ear, or a paffage for the mucus of thefe parts; the external opening having a particular form would incline us to believe, that fomething was conveyed to the tympanum.

The bony part of the organ is very hard and brittle, rendering it even difficult to be cut with a faw, without its chipping into pieces. That part which contains the immediate organ is by much the hardeft, and has a very fmall portion of animal fubftance in it; for when fteeped in an acid, what remains is very foft, almoft like a jelly, and laminated. The bone is not only harder in its fubftance, but there is on the whole more folid bone than in the correfponding parts of quadrupeds, it being thick and maffy.

The part containing the tympanum is a thin bone, coiled upon itfelf, attached by one end to the portion which contains the organ; and this attachment in fome is by clofe contact only, as in the Narwhale; in others, the bones run into one another, as in the Bottle-nofe and Piked Whales.

The concave fide of the tympanum is turned towards the organ, its two edges being clofe to it; the outer is irregular, and in many only in contact, as in the Porpoife: while in

* Thefe communications with the Euftachian tube may be compared to a large bag on the bafes of the fkull of the Horfe and Afs, which is a lateral fwell of the membranous part of the tube, and when diftended will contain nearly a quart.

others

others the union is by bony continuity, as in the **Bottle-nose**
Whale, leaving a paſſage on which the membrana tympani
is ſtretched, and another opening, which is the communica-
tion with the ſinuſes.

The ſurface of the bone containing the immediate organ oppoſite to the mouth of the tympanum is very irregular, having a
number of eminences and cavities. The cavity of the tympanum is lined with a membrane, which alſo covers the ſmall
bones with their muſcles, and appears to have a thin cuticle.
This membrane renders the bones, muſcles, tendons, &c.
very obſcure, which are ſeen diſtinctly when that is removed.
It appears to be a continuation of the perioſteum, and the
only uniting ſubſtance between the ſmall bones. Beſides
the general lining, there is a plexus of veſſels, which is
thin and rather broad, and attached by one edge, the reſt being
looſe in the cavity of the tympanum, ſomewhat like the plexus
choroides in the ventricles of the brain. The cavity, we may
ſuppoſe, intended to increaſe ſound, probably by the vibration
of the bone; and from its particular formation we can eaſily conceive, that the vibrations are conducted, or reflected, towards
the immediate organ, it being in ſome degree a ſubſtitute for
the external ear.

The external opening being ſmaller than in any animals of
the ſame ſize, the membrana tympani is nearly in the ſame
proportion. In the Bottle-noſe Whale, the Grampus, and
Porpoiſe, it is ſmooth and concave externally, but of a particular conſtruction on the inner ſurface; for a tendinous pro-
ceſs paſſes from it towards the malleus, converging as it pro-
ceeds from the membrane, and becoming thinner till its inſertion
into that bone. I could not diſcover whether it had any muſcular fibres which could affect the action of the malleus. In

the

the Piked Whale, the termination of the external opening, inftead of being fmooth and concave, is projecting, and returns back into the meatus for above an inch in length, is firm in texture, with thick coats, is hollow on its infide, and its mouth communicating with the tympanum ; one fide being fixed to the malleus, fimilar to the tendinous procefs which goes from the infide of the membrana tympani in the others.

A little way within the membrana tympani, are placed the fmall bones, which are three in number, as in the quadruped, Malleus, Incus, and Stapes ; but in the Bottle-nofe Whale there is a fourth, placed on the tendon of the Stapedæus mufcle. Thefe bones are as it were fufpended between the bone of the tympanum, and that of the immediate organ.

The malleus has two attachments, befides that with the incus ; one clofe to the bone of the tympanum, which, in the Porpoife, is only by contact, but in others by a bony union ; the other attachment is formed by the tendon, above defcribed, being united to the inner furface of the membrana tympani. Its bafe articulates with the incus.

The incus is attached by a fmall procefs to the tympanum, and is fufpended between the malleus and ftapes. The procefs by which it articulates with the ftapes is bent towards that

The ftapes ftands on the veftibulum, by a broad oval bafe. In many of this tribe, the opening from fide to fide of the ftapes is fo fmall as hardly to give the idea of a ftirrup.

The mufcles which move thefe bones are two in number, and tolerably ftrong. One arifes from that projecting part of the tympanum which goes to form the Euftachian tube, and running backwards is inferted into a fmall depreffion on the anterior part of the malleus. The ufe of this mufcle feems to

5   be

be to tighten the membrana tympani; but in thofe which have
the malleus anchylofed with the tympanum, we can hardly
conjecture its ufe. The other has its origin from the inner
furface of the tympanum, and paffing backwards is inferted
into the ftapes by a tendon, in which I found a bone in the
large Bottle nofe. This mufcle gives the ftapes a lateral motion. What particular ufe in hearing may be produced by
the action of thefe mufcles, I will not pretend to fay; but we
muft fuppofe, whatever motion is given to the bones muft
terminate in the movement of the ftapes.

The immediate organ of hearing is contained in a round,
bony procefs, and confifts of the Cochlea and Semicircular
Canals, which fomewhat refemble the quadruped; but, befides
the two fpiral turns of the cochlea, there is a third, which
makes a ridge within that continued from the foramen rotundum, and follows the turns of the canal.

The cochlea is much larger, when compared with the femicircular canals, than in the human fpecies and quadruped.

We may reckon two paffages into the immediate organ of hearing, the foramen rotundum, and foramen ovale. They are at a
greater diftance than in the quadruped. The foramen rotundum
is placed much more on the outer furface of the bone, and not
in the cavity of the bony tympanum; but may be faid to communicate with the furrounding cellular part of the tympanum.
The foramen rotundum, which is the beginning of one of
thefe turns, appears to be only one end of a tranfverfe
groove, which is afterwards clofed in the middle, forming a
canal with the two ends open; fo that this foramen appears
to have two beginnings; but the other opening is probably
only a paffage for blood-veffels going to the cochlea.

From

From this foramen begins the inner turn of the cochlea, which is the largeft, efpecially at its beginning; the other begins from the veftibulum. The cochlea is a fpiral canal coiled within itfelf, and divided into two by a thin fpiral bony plate, which is compleated in the recent fubject, and forms two perfect canals.

In the recent fubject, the foramen rotundum is lined with the membrane of the tympanum, which terminates in a blind end, forming a kind of membrana cochleæ. The other opening, in the recent fubject, communicates with the fpiral turn, beyond the membranous termination of the foramen rotundum.

The foramen ovale has a little projection inwards all round, on which the ftapes ftands: within this is the veftibulum, which is common to the other fpiral turn of the cochleæ, and the femicircular canals; this canal of the cochlea paffes out firft in a direction contrary to its general courfe, but foon makes a turn into the fpiral. It is round, and not merely a divifion of the cochlea into two by a feptum, but has a membrane of its own, which is attached to the thin bony plate, and lines that part of the cochlea in fuch a manner as to retain its ftructure when the bone is removed. The cochlea in fome compleats one turn and an half; in others, more. It is not a fpiral on a plane, or cylinder, but on a cone.

I have already obferved, that by looking in at the foramen rotundum, we fee two fmall ridges; the uppermoft is the fwell of the canal from the veftibulum juft defcribed; the lower ridge, which is alfo a canal, may be obferved juft to pafs along the foramen belonging to this canal, clofe to the feptum between the two; a circumftance, I believe, peculiar to this tribe. Its beginning is clofe to the veftibulum, but does not open from it, and paffes along the firft defcribed fpiral

6   turn

turn to its apex: when opened, it appears to be a canal full of fmall perforations, probably the paffages of the branches from the auditory nerve.

This bony procefs has feveral perforations in it ; one of them large, for the paffage of the feventh pair of nerves. The fize of the portio mollis, before its entrance into the organ, is very large, and bears no proportion to that which enters. The paffage for this nerve is very wide, and feems to have an irregular blind conical, and fomewhat fpiral, termination; its being fpiral arifes from the clofenefs to the point of the cochlea.

In the terminating part there are a number of perforations into the cochlea, and one into the femicircular canals, which afford a paffage to the different divifions of the auditory nerve. There is a confiderable foramen in its anterior fide near the bottom, for the paffage of the portio dura, and which is continued backward to the cavity of the tympanum near the ftapes, and emerges near the pofterior and upper part of this bone.

Of the Organ of Seeing.

The eye in this tribe of animals is conftructed upon nearly the fame principle as that of quadrupeds, differing, however, in fome circumftances; by which it is probably better adapted to fee in the medium through which the light is to pafs. It is upon the whole fmall for the fize of the animal, which would lead to the fuppofition, that their locomotion is not great; for, I believe, animals that fwim are in this refpect fimilar to thofe that fly; and as this tribe come to the furface of the medium in which they live, they may be confidered in the fame view with birds which foar; and we find, birds that

that fly to great heights, and move through a confiderable
fpace, in fearch of food, have their eyes larger in proportion
to their fize.

The eyelids have but little motion, and do not confift of
loofe cellular membrane, as in quadrupeds, but rather of the common adipofe membrane of the body; the connexion, however,
of their circumference with the common integuments is
loofe, the cellular membrane being lefs loaded with oil, which
allows of a flight fold being made upon the furrounding parts
in opening the eyelids. This is not to an equal degree in them
all, being lefs fo in the Porpoife than in the Piked Whale.

The tunica conjunctiva, where it is reflected from the eyelid
to the eyeball, is perforated all round by fmall orifices of the
ducts of a circle of glandular bodies lying behind it.

The lachrymal gland is fmall; its ufe being fupplied by thofe
above-mentioned; and the fecretion from them all, I believe,
to be a mucus fimilar to what is found in the Turtle and Cro-
codile. There are neither puncta nor lachrymal duct, fo that
the fecretion, whatever it be, is wafhed off into the water.

The mufcles which open the eyelids are very ftrong: they
take their origin from the head, round the optic nerve, which
in fome requires their being very long, and are fo broad as
almoft to make one circular mufcle round the whole of the
interior ftraight mufcles of the eye itfelf. They may be divided into four; a fuperior, an inferior, and one at each
angle: as they pafs outwards to the eyelids, they diverge and
become broader, and are inferted into the infide of the eyelids
almoft equally all round. They may be termed the dilatores of
the eyelids; and, before they reach their infertion, give off
the external ftraight mufcles, which are fmall, and inferted
into the fclerotic coat before the tranfverfe axis of the eye:
thefe

thefe may be named the elevator, depreffor, adductor, and ab-
ductor, and may be diffected away from the others as diftinct
mufcles. Befides thefe four going from the mufcles of the
eyelid to the eye itfelf, there are two which are larger, and inclofe the optic nerve with the plexus. As thefe pafs outwards
they become broad, may in fome be divided into four, and
are inferted into the fclerotic coat, almoft all round the eye,
rather behind its tranfverfe axis.

The two oblique mufcles are very long; they pafs through
the mufcles of the eyelids, are continued on to the globe of
the eye, between the two fets of ftraight mufcles, and at
their infertions are very broad; a circumftance which gives
great variation to the motion of the eye.

The fclerotic coat gives fhape to the eye, both externally
and internally, as in other animals; but the external fhape and
that of the internal cavity are very diffimilar, arifing from
the great difference in the thicknefs of this coat in different
parts. The external figure is round, except that it is a little
flattened forwards; but that of the cavity is far otherwife,
being made up of fections of various circles, being a little
lengthened from the inner fide to the outer, a tranfverfe fection
making a fhort ellipfis.

In the Piked Whale the long axis is two inches and three-
quarters, the fhort axis two inches and one-eighth.

The pofterior part of the cavity is a tolerably regular curve,
anfwering to the difference in the two axifes; but forwards,
near the cornea, the fclerotic coat turns quickly in, to meet

rotic coat and the bottom of the eye not above an inch and a quarter.

In the Piked Whale the ſclerotic coat, at its poſterior part, is very thick: near the extreme of the ſhort axis it was half an inch, and at the long axis one-eighth of an inch thick. In the Bottle-noſe Whale, the extreme of the ſhort axis was half an inch thick, and the extremes of the long axis about a quarter of an inch, or half the other.

The ſclerotic coat becomes thinner as it approaches to its union with the cornea, where it is thin and ſoft. It is extremely firm in its texture, where thick, and from a tranſverſe ſection would ſeem to be compoſed of tendinous fibres, intermixed with ſomething like cartilage; in this ſection four paſſages for veſſels remain open. This firmneſs of texture precludes all effect of the ſtraight muſcles on the globe of the eye, by altering its ſhape, and adapting its focus to different diſtances of objects, as has been ſuppoſed to be the caſe in the human eye.

The cornea makes rather a longer ellipſis than the ball of the eye; the ſides of which are not equally curved, the upper being moſt conſiderably ſo. It is a ſegment of a circle ſomewhat ſmaller than that of the eyeball, is ſoft and very flaccid.

The tunica choroides reſembles that of the quadruped; and its inner ſurface is of a ſilver hue, without any nigrum pigmentum.

The nigrum pigmentum only covers the ciliary proceſſes, and lines the inſide of the iris.

The retina appears to be nearly ſimilar to that of the quadruped.

The arteries going to the coats of the eye form a plexus paſſing round the optic nerve, reſembling, in its appearance, that of the ſpermatic artery in the Bull and ſome other animals.

The

The cryftalline humor refembles that of the quadruped; but whether it is very convex or flattened, I cannot determine, thofe I have examined having been kept too long to preferve their exact fhape and fize. The vitreous humor adhered to the retina at the entrance of the optic nerve.

The optic nerve is very long in fome fpecies, owing to the vaft width of the head.

I fhall not at prefent confider the eye in animals of this tribe, as it refpects the power of vifion, that being performed on a general principle common to every animal inhabiting the water; more efpecially as I am only mafter of the conftruction and formation of the eye, and not of the fize, fhape, and denfities of the humors; yet, from reafoning, we muft fu pofe them to correfpond with the fhape of the eye, and the medium through which the light is to pafs.

Of the Parts of Generation.

The parts of generation in both fexes of this order of animals come nearer in form to thofe of the ruminating than of any others; and this fimilarity is, perhaps, more remarkable in the female than in the male; for their fituation in the male muft vary on account of external form, as was before obferved.

The tefticles retain the fituation in which they were formed, as in thofe quadrupeds in which they never come down into the fcrotum. They are fituated near the lower part of the abdomen, one on each fide, upon the two great depreffors of the tail. At this part of the abdomen, the tefticles come in contact with the abdominal mufcles anteriorly.

The vafa deferentia pafs directly from the epididymis behind the bladder, or between it and the rectum, into the urethra;

Qqq2 and

and there are no bags fimilar to thofe called veficulæ feminales
in certain other animals.

The ſtructure of the penis is nearly the fame in them all,
and formed much upon the principle of the quadruped. It is
made up of two crura, uniting into one corpus cavernofum, and
the corpus fpongiofum feems firſt to enter the corpus cavernofum. In the Porpoife, at leaſt, the urethra is found nearly
in the center of the corpus cavernofum; but towards the
glans feems to feparate or emerge from it, and becoming a
diſtinct fpongy body, runs along its under furface, as in quadrupeds. The corpus cavernofum in fome is broader from the
upper part to the lower than from fide to fide; but in the Porpoife has the appearance of being round, becoming fmaller forwards, fo as to terminate almoſt in a point fome diſtance from
the end of the penis. The glans does not fpread out as in many
quadrupeds, but feems to be merely a plexus of veins covering
the anterior end of the penis, yet is extended a good way further on, and is in fome no more than one vein deep.

The crura penis are attached to two bones, which are nearly
in the fame fituation and in the fame part of the pelvis as thofe
to which the penis is attached in quadrupeds; but thefe bones
are only for the infertion of the crura, and not for the fupport of
any other part, like the pelvis in thofe animals which have
poſterior extremities, neither do they meet at the fore part, or
join the vertebræ of the back.

The erectores penis are very ſtrong mufcles, having an origin
and infertion fimilar to thofe of the human fubject.

The acceleratores mufcles are likewife very ſtrong; and there
is a ſtrong and long mufcle, arifing from the anus, and paffing
forwards to the bulb of the penis, that runs along the under
furface of the urethra, and is at laſt loſt or inferted in the cor-

pus

pus fpongiofum. This mufcle draws the penis into the prepuce, and throws that part of the penis that is behind its infertion into a ferpentine form. It is common to moft animals that draw back the penis into what is called the fheath,
and may be called the retractor penis.

In all the females which I have examined, the parts of
generation are very uniformly the fame; confifting of the
external opening, the vagina, the two horns of the uterus,
Fallopian tubes, fimbriæ, and ovaria.

The external opening is a longitudinal flit, or oblong opening, whofe edges meet in two oppofite points, and the fides
are rounded off, fo as to form a kind of fulcus. The fkin and
parts on each fide of this fulcus are of a loofer texture than on
the common furface of the animal, not being loaded with oil,
and allowing of fuch motion of one part on another as admits
of dilatation and contraction. The vagina paffes upwards and
backwards towards the loins, fo that its direction is diagonal
refpecting the cavity of the abdomen, and then divides into the
two horns, one on each fide of the loins; thefe afterwards
terminating in the Fallopian tubes, to which the ovaria are
attached. From each ovarium there is a fmall fold of the
peritoneum, which paffes up towards the kidney of the fame
fide, as in moft quadrupeds.

The infide of the vagina is fmooth for about one-half of its
length, and then begins to form fomething fimilar to valves
projecting towards the mouth of the vagina, each like an os
tincæ; thefe are about fix, feven, eight, or nine in number.
Where they begin to form, they hardly go quite round, but
the laft are complete circles. At this part too the vagina becomes fmaller, and gradually decreafes in width to its termination. From the laft projecting part, the paffage is continued

nued up to the opening of the two horns, and the inner surface of this laft part is thrown into longitudinal rugæ, which are continued into the horns. Whether this laft part is to be reckoned common uterus or vagina, and that the laft valvular part is to be confidered as os tincæ, I do not know; but from its having the longitudinal rugæ, I am inclined to think it is uterus, this ftructure appearing to be intended for diftinction.

The horns are an equal divifion of this part; they make a gentle turn outwards, and are of confiderable length. Their inner furface is thrown into longitudinal rugæ, without any fmall protuberances for the cotyledons to form upon, as in thofe of ruminating animals; and where they terminate, the Fallopian tubes begin.

In the Bottle-nofe Whale, where the Fallopian tubes opened into the horns of the uterus, they were furrounded by pendulous bodies hanging loofe in the horns.

The Fallopian tubes, at their termination in the uterus, are remarkably fmall for fome inches, and then begin to dilate rather fuddenly; and the nearer to the mouth the more this dilatation increafes, like the mouth of a French horn, the termination of which is five or fix inches in diameter. They are very full of longitudinal rugæ through their whole length.

The ovaria are oblong bodies, about five inches in length; one end attached to the mouth of the Fallopian tube, and the other near to the horn of the uterus. They are irregular on their external furface, refembling a capfula renalis or pancreas. They have no capfula, but what is formed by the long Fallopian tube.

How the male and female copulate, I do not know; but it is alledged, that their pofition in the water is erect at that time, which I can readily fuppofe may be true; for otherwife,

if

if the connexion is long, it would interfere with the act of respiration, as in any other position the upper surface of the heads of both could not be at the surface of the water at the same time. However, as in the parts of generation they most resemble those of the ruminating kind, it is possible they may likewise resemble them in the duration of the act of copulation; for, I believe, all the ruminants are quick in this act.

Of their uterine gestation I as yet know nothing; but it is very probable, that they have only a single young one at a time, there being only two nipples. This seemed to be the case with the Bottle-nose Whale, caught near Berkeley, which had been seen for some days with one young one following it, and they were both caught together.

The glands for the secretion of milk are two; one on each side of the middle line of the belly at its lower part. The posterior ends, from which go out the nipples, are on each side of the opening of the vagina, in small sulci. They are flat bodies lying between the external layer of fat and abdominal muscles, and are of considerable length, but only one-fourth of that in breadth. They are thin, that they may not vary the external shape of the animal, and have a principal duct, running in the middle through the whole length of the gland, and collecting the smaller lateral ducts, which are made up of those still smaller. Some of these lateral branches enter the common trunk in the direction of the milk's passage, others in the contrary direction, especially those nearest to the termination of the trunk in the nipple. The trunk is large, and appears to serve as a reservoir for the milk, and terminates externally in a projection, which is the nipple. The lateral portions of the fulcus which incloses the nipple, are composed of parts looser in texture than the common adipose membrane, which is probably to admit of the elongation or projection of

5   the

the nipple. On the outſide of this there is another ſmall
fiſſure, which, I imagine, is likewiſe intended to give greater
facility to the movements of all theſe parts.

The milk is probably very rich; for in that caught near
Berkeley with its young one, the milk, which was taſted by
Mr. Jenner and Mr. Ludlow, Surgeon, at Sodbury, was
rich like Cow's milk to which cream had been added.

The mode in which theſe animals muſt fuck would appear
to be very inconvenient for reſpiration, as either the mother or
young one will be prevented from breathing at the time, their
noſtrils being in oppoſite directions, therefore the noſe of one
muſt be under water, and the time of fucking can only be between each reſpiration. The act of fucking muſt likewiſe be
different from that of land animals; as in them it is performed
by the lungs drawing the air from the mouth backwards into
themſelves, which the fluid follows, by being forced into the
mouth from the preſſure of the external air on its ſurface;
but in this tribe, the lungs having no connexion with the
mouth, fucking muſt be performed by ſome action of the
mouth itſelf, and by its having the power of expanſion.

Pl. 6. 1.

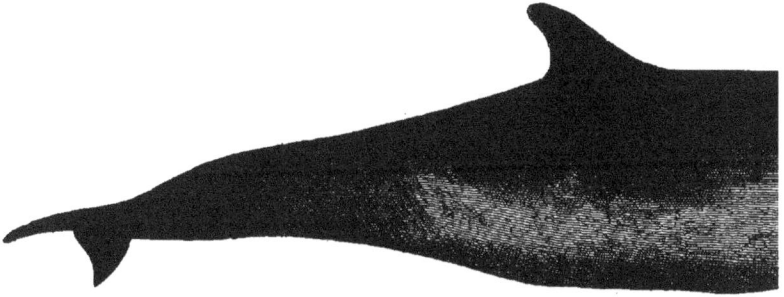

Fig. 1.

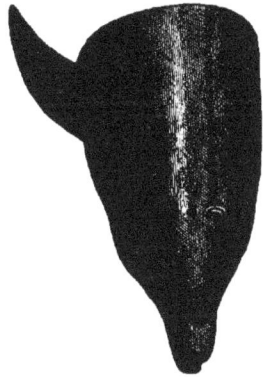

Fig.

Fig. 2.

Basire sc.

Basire Sc.

	F.	I.
Whole Length...................	17	0
Upper Jaw from Eye to Eye......1	:	8
Lower Jaw......................2	:	6
Within the Whalebone 0	:	10½
Greatest length of Whalebone ..0	:	5

Fig

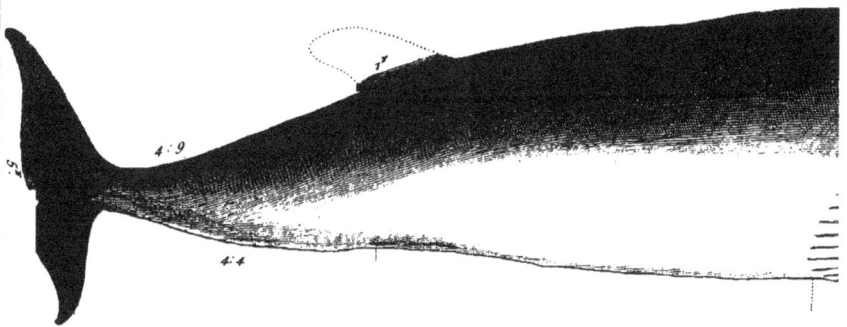

1″

4 : 9

5

4 : 4

Fig. 2.

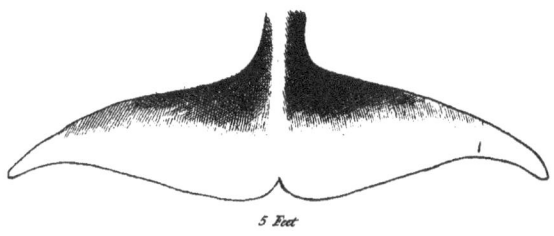

5 Feet

Balæn

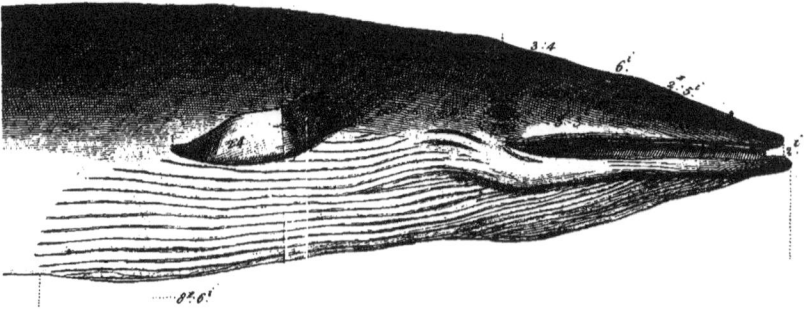

Fig. 1.

Balæna Rostrata.
Fabricius.

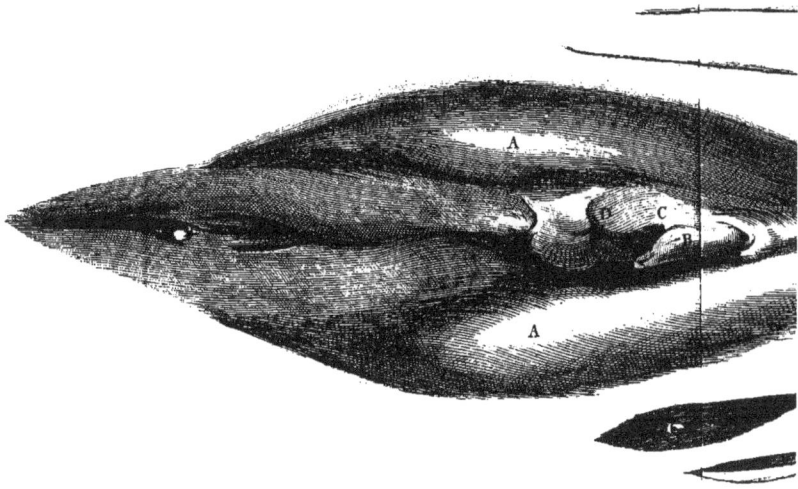

Feet inches

1   3

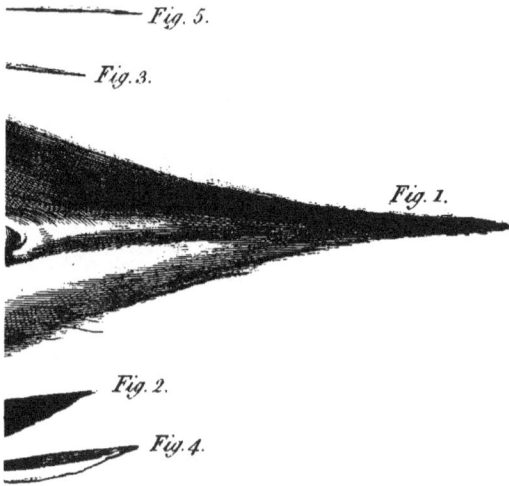

Fig. 5.

Fig. 3.

Fig. 1.

Fig. 2.

Fig. 4.

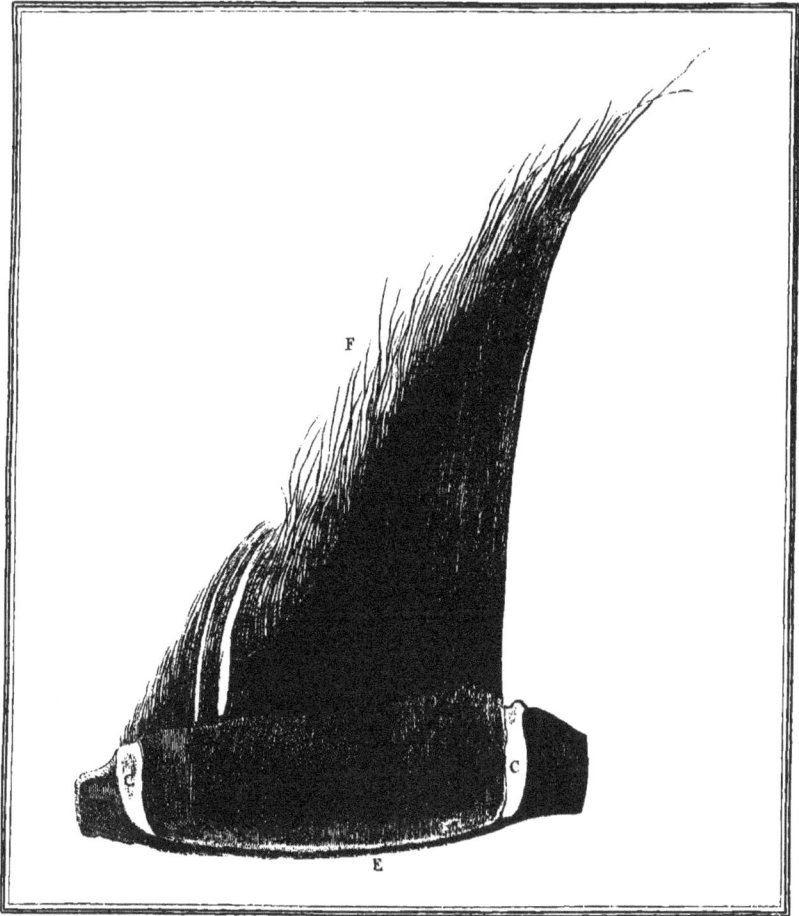

F

C

E

Fig. 1.

Fig. 2.

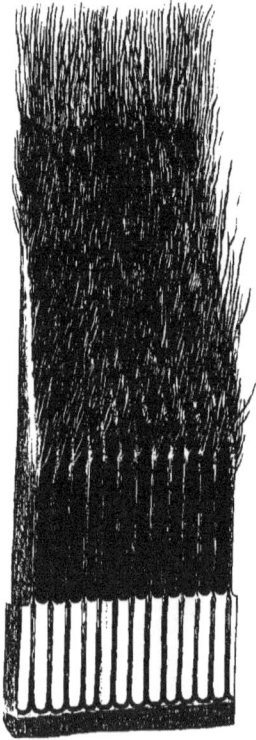

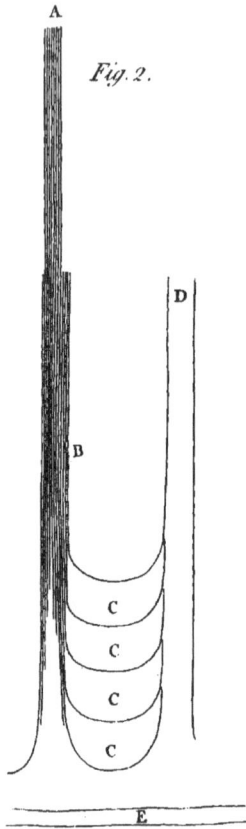

EXPLANATION OF THE PLATES.

PLATE XVI.

This fish is called a Grampus: it was caught in the mouth of the river Thames, in the year 1759, and brought up to Westminster Bridge in a barge. It was twenty-four feet long.

PLATE XVII.

Another species of Grampus, which was caught in the river Thames, fifteen years ago. It was eighteen feet long.

PLATE XVIII.

Fig. 1. A species of Bottle-nose Whale; the Delphinus Delphis of LINNÆUS. It was caught upon the sea-coast, near Berkeley, where it had been seen for several days, following its mother, and was taken along with the old one, and sent up to me whole, for examination, by Mr. JENNER, Surgeon, at Berkeley. The old one was eleven feet long.

Fig. 2. The head of the same Whale as fig. 1. to shew the shape of the blow-hole, which is transverse, and almost semi-circular.

PLATE XIX.

The Bottle-nose Whale described by DALE. It is similar to that of Plate XVIII. in its general form, but has only two small pointed teeth in the fore part of the upper jaw, and is rather lighter coloured on the belly. It was caught above

London Bridge in the year 1783, and became the property of the late Mr. Alderman P<small>UGH</small>, who very politely allowed me to examine its ſtructure, and to take away the bones. It was twenty-one feet long.

PLATE XX.

Fig. 1. The Balæna Roſtrata of F<small>ABRICIUS</small>, or Piked Whale. It was caught upon the Dogger Bank. It had met with ſome accident between the two lower jaws under the tongue, in which part a conſiderable collection of air had taken place, ſo as to raiſe up the tongue and its attachments into a round body in the mouth, projecting even beyond the jaws. This ren-dered the head ſpecifically lighter than the water, ſo that it could not ſink, and therefore was eaſily caught.

It was ſeventeen feet long, and was brought to St. George's Fields, where I purchaſed it. The dorſal fin having been cut off cloſe to the back, is therefore only marked by a dotted line. Fig. 2. A view of the tail, to ſhew its breadth.

PLATE XXI.

Includes the external parts of generation, with the relative ſituation of the anus and the nipples, of the Balæna Roſtrata.

Fig. 1. The labia pudendi ſpread open, expoſing the meatus urinarius, vagina, and anus, which in a natural ſtate are all concealed, there only appearing a long ſlit, the two edges of which are in contact.

AA. The labia pudendi.
B. The clitoris.
C. The meatus urinarius.

D. The

D. The opening of the vagina.

E. The anus.

Fig. 2. The fulcus, in which the left nipple lies, fpread open, and the nipple itfelf expofed to view.

Fig. 3. The fulcus of the right nipple, in a natural ftate, only appearing like a line.

Fig. 4. A fulcus near to the nipple, which is fpread open to fhew the infide. This fulcus, I conceive, gives a freedom to the motion of the fkin of thefe parts, fo as to allow the nipple to be more freely expofed.

Fig. 5. The fame fulcus on the oppofite fide, clofed up.

PLATE XXII.

A fide view of one of the plates of whalebone of the Balæna Roftrata.

A. The part of the plate which projects beyond the gum.

B. The portion which is funk into the gum.

CC. A white fubftance, which furrounds the whalebone, forming there a projecting bead, and alfo paffing between the plates, to form their external lamellæ.

DD. The part analogous to the gum.

E. A flefhy fubftance, covering the jaw bone, and on which the inner lamella of the plate is formed.

F. The termination of the plate of whalebone in a kind of hair.

PLATE XXIII.

Fig. 1. A perpendicular fection of feveral plates of whalebone in their natural fituation in the gum; their inner edges, or fhorteft terminations, are removed, and the cut edges of the plates feen from the infide of the mouth.

The

The upper part fhews the rough furface formed by the hairy termination of each plate of whalebone.

The middle part fhews the diftance the plates of whalebone are from each other.

The lower part fhews the white fubftance in which they grow, and alfo the bafis on which they ftand.

Fig. 2. An outline confiderably magnified, to fhew the mode of growth of the plates, and of the white intermediate fubftance.

A. The middle layer of the plate, which is formed upon a pulp or cone that paffes up in the centre of the plate. The termination of this layer forms the hair.

B. One of the outer layers, which grows, or is formed, from the intermediate white fubftance.

CCCC. The intermediate white fubftance, laminæ of which are continued along the middle layer, and form the fubftance of the plate of whalebone.

D. The outline of another plate of whalebone.

E. The bafis on which the plates are formed, which adheres to the jaw bone.

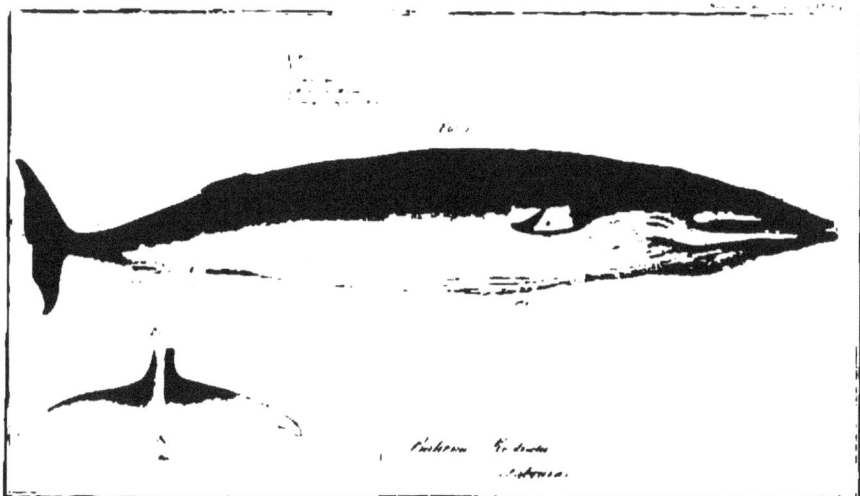

Balæna rostrata
....

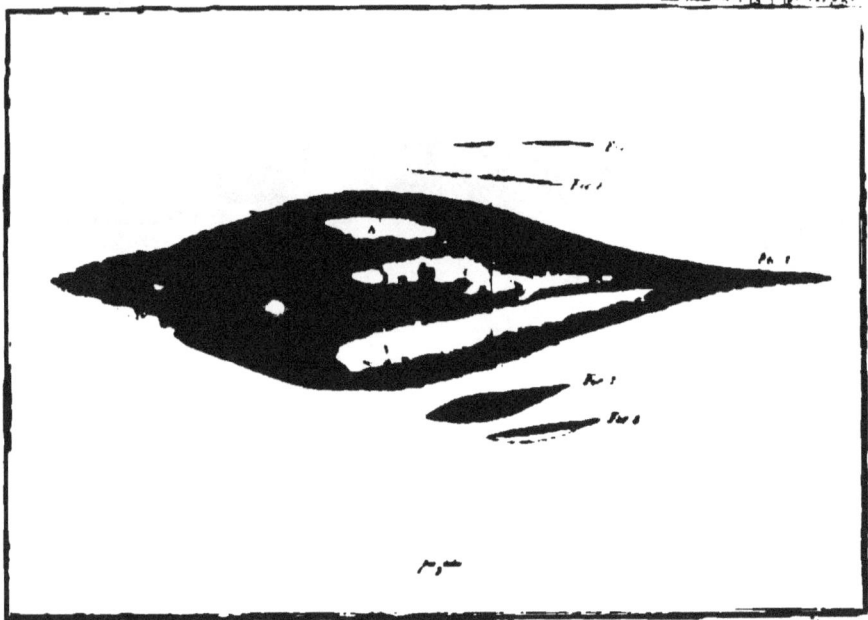

www.ingramcontent.com/pod-product-compliance
Lightning Source LLC
Chambersburg PA
CBHW021946190326
41519CB00009B/1153